Cerium - Chemistry, Technology, Geology, Soil Science and Economics

Edited by Michael Aide

Published in London, United Kingdom

Cerium - Chemistry, Technology, Geology, Soil Science and Economics
http://dx.doi.org/10.5772/intechopen.110944
Edited by Michael Aide

Contributors
A. Manikandan, Christian Girginov, Kannan Elaya Kumar, Leelakhrishna Reddy, Machaba L.A. Letswalo, Michael Aide, Mousri Paul, Plamen Petkov, S. Muthulingam, Sabyasachi Karmakar, Stephan Kozhukharov, Supratic Chakraborty, Vanya Lilova

First published in London, United Kingdom, 2024 by IntechOpen
IntechOpen is the global imprint of INTECHOPEN LIMITED, registered in England and Wales, registration number: 11086078, 167-169 Great Portland Street, London, W1W 5PF, United Kingdom

British Library Cataloguing-in-Publication Data
A catalogue record for this book is available from the British Library

Additional hard and PDF copies can be obtained from orders@intechopen.com

Cerium - Chemistry, Technology, Geology, Soil Science and Economics
Edited by Michael Aide
p. cm.
Print ISBN 978-0-85466-200-5
Online ISBN 978-0-85466-199-2
eBook (PDF) ISBN 978-0-85466-201-2

For EU product safety concerns:
IN TECH d.o.o., Prolaz Marije Krucifikse Kozulić 3, 51000 Rijeka, Croatia,
info@intechopen.com or visit our website at intechopen.com.

Meet the editor

Dr. Michael Aide received a bachelor's degree in Chemistry and Mathematics from the University Wisconsin-Madison, USA, and a Ph.D. in Soil Chemistry from Mississippi State University, USA. He has been an educator and agronomic-soil researcher at Southeast Missouri State University, USA, since 1982. His research interests include the growth and development of rice in integrated systems involving soil fertility, water management, and mitigation of arsenic uptake. Rice research has permitted his travel to southeastern Asia, Central America, and Egypt. Dr. Aide is a certified soil scientist who has investigated rare earth elements in soils, attempting to utilize them to indicate parent material uniformity and the fate/transport. His professional affiliations include the Soil Science Society America and the Soil and Water Conservation Society.

Contents

Preface

Cerium – Chemistry, Technology, Geology, Soil Science and Economics presents a compelling array of research detailing potential new mining resources and aspects of the technology and utilization of rare earth elements. The first section discusses the geology and soil science of ion-exchange rare earth element deposits, which offers a unique resource for procurement and provides an understanding of their formation and chemical properties. The second section discusses cerium compounds in LED technology, the effect of silver nanoclusters on the physical and electrical properties of cerium oxide thin film, the impact of rare earth element cerium on structural and magnetic characterization of different nanoferrites, and innovative applications of cerium oxide-based materials in civil engineering, automation, and energy sectors. Each chapter provides a unique perspective on the uses or technology associated with rare earth elements, especially cerium.

Dr. Michael Aide
Professor of Agriculture,
Southeast Missouri State University,
Cape Girardeau, MO, USA

Section 1

The Geology and Soil Science of the Cerium

Chapter 1

The Soil Chemistry of Cerium with an Emphasis on the Formation of Ion-Adsorption Rare Earth Element Deposits

Michael Aide

Abstract

Cerium is an important rare earth element that has numerous and essential technological applications, as do many of the rare earth elements. Given that the rare earth elements do not exist as abundant and highly concentrated ore deposits, considerable research has been devoted toward their economically feasible extraction and subsequent processing. Ion-adsorption rare earth element deposits are emerging sources for rare earth element extraction and processing, including cerium. This manuscript presents a brief introduction to the soil thermodynamics of rare earth element hydrolysis, complexation, and adsorption onto phyllosilicates. These intrinsically critical thermodynamic-based activities govern rare earth element mineral weathering, species mobility, bioavailability, and suitability for specific extraction protocols. Ion-adsorption rare earth element deposits and their formation are discussed to provide options for subsequent research involving resource utilization, conservation, and environmental protection.

Keywords: rare earth elements, ion-adsorption type deposits, Ce hydrolysis, REE minerals, weathering

1. Introduction

The rare earth elements (REE's) are cerium (Ce), praseodymium (Pr), neodymium (Nd), promethium (Pm), samarium (Sm), europium (Eu), gadolinium (Gd), terbium (Tb), dysprosium (Dy), holmium (Ho), erbium (Er), thulium (Tm), ytterbium (Yb), and lutetium (Lu). Frequently, lanthanum (La) is included because of its position in the periodic table and its trivalent oxidation state. Yttrium (Y), a transition metal with atomic number 39, has an electronic configuration of [Kr] $4d^1 5s^2$. Yttrium is frequently included with the rare earth elements because of its chemical similarity, and its natural occurrence is closely aligned with that of the heavy rare earth elements. Promethium's presence in the natural environment is virtually nonexistent because of radioactive decay (half-life is 2.62 years) [1]. The light rare earth elements range from La to Eu, whereas the heavy rare earth elements range from Gd to Lu.

The Lanthanide series has ground state electronic configurations with the 4f electronic orbitals having at least one electron. Given cerium with an atomic number of 58, the corresponding electron configuration is [Xe] $4f^1\ 5d^1\ 6s^2$, and the electron configuration for the trivalent species (Ce^{3+}) is [Xe] $4f^1$. The Ce^{4+} species has an electron configuration of [Xe]. The lanthanide contraction is attributed to the reduced ability of the inner-shell electrons to shield the outer-shell electrons; thus, with an increasingly charged nucleus, the greater electron attraction affords a decreasing atomic radius [1]. **Figure 1** illustrates the progressive reduction of the atomic radii with increasing atomic number for lanthanum and the rare earth elements.

The rare earth elemental distributions vary across igneous, metamorphic, and sedimentary rocks, features attributed to diverse mineral compositions. Considering sedimentary rocks, argillaceous rocks typically exhibit greater rare earth abundances than sandstones; however, the variances are sufficiently large to limit comparisons (**Figure 2**). Similarly, sandstones typically have greater rare earth abundances than carbonaceous rocks [2]. For all these rock types, the light REE distribution typically exhibit greater elemental abundances than the heavy REE. The light and heavy REE's typically reside in discretely different mineral assemblages.

Numerous minerals have cerium and other rare earth elements as lattice constituents. Some of the more abundant Ce-bearing minerals include (i) cerianite (CeO_2), (ii) fluorite (CaF with Ca replacement with Y and Ce), (iii) allanite ((CaCe)($AlAlFe^{2+}$)O[Si_2O_7][SiO_4](OH)), (iv) apatite ($Ca_5(PO_4)3(Cl,OH,F)$ with REE substitution for Ca), (v) monazite (Ce,La,Nd,Th)(PO_4,SiO_4), (vi) xenotime (YPO_4 with Y substitution with Dy, Er, Tb, and Yb), (vii) rhabdophane ((CeLa) PO_4), and (viii) bastnaesite (CaREE which is a fluorocarbonate) [2–4]. Many rare earth element-bearing minerals occur because of isomorphic substitution during magma cooling.

The Portageville soil (fine, smectitic, calcareous, thermic Vertic Endoaquolls) is a clayey textured soil formed in Mississippi River alluvium. The rare earth element distribution of this soil series reflects the Oddon-Harkins rule, where for each element with an even atomic number, the next element in the periodic table, with an

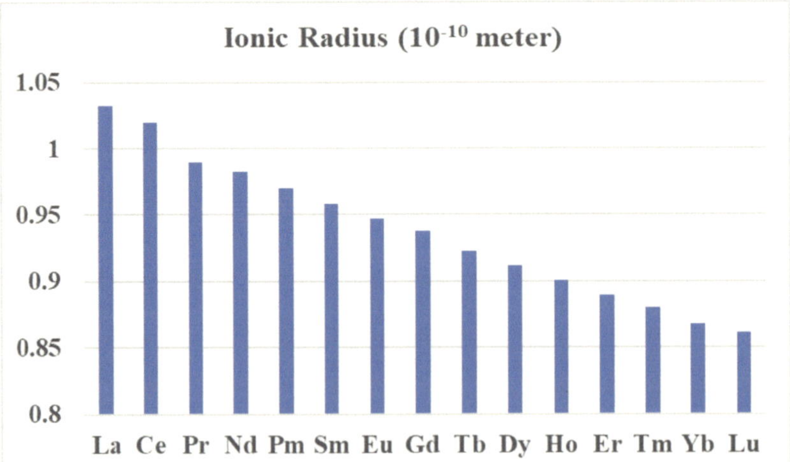

Figure 1.
The ionic radius for six coordination of the rare earth elements.

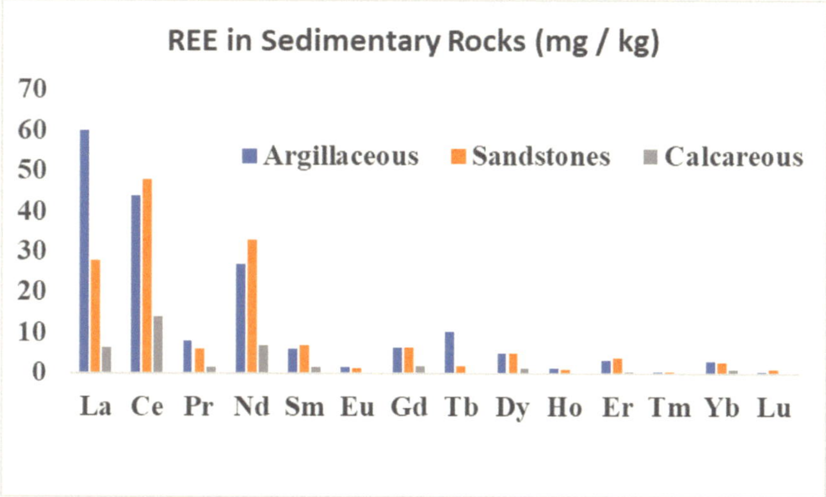

Figure 2.
The comparison of the rare earth element abundances for argillaceous, sandstones, and calcareous rocks (source [3]).

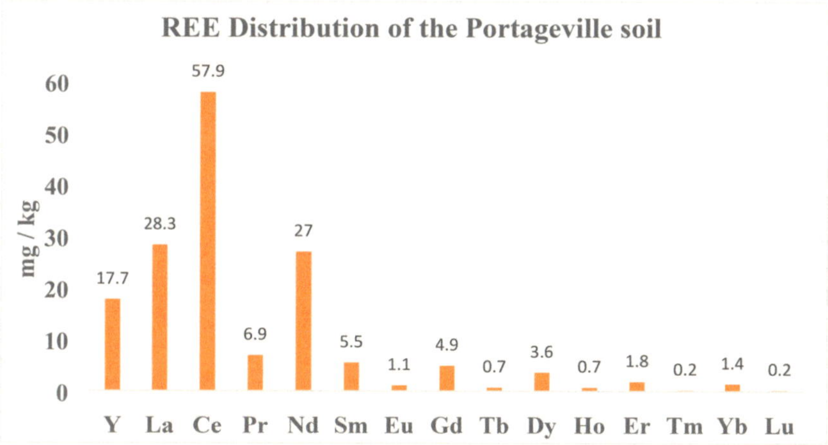

Figure 3.
The rare earth element distributions of and a clayey textured soil (Portageville).

odd atomic number, generally has a smaller abundance. The light REEs typically are exceedingly more abundant than the heavy REEs (**Figure 3**).

Many nations have rare earth element mining operations; however, several nations dominate the rare earth element production (**Table 1**). China is currently the world's leading producer with 240,000 metric tons in 2023 and substantial proven and indicated reserves [5]. Brazil, Canada, and Vietnam have reserve capability to substantially increase production [5].

The objectives of this manuscript are to provide a (i) brief review of rare earth element chemical and geological properties, (ii) review soil chemical properties such as hydrolysis, complexation and adsorption onto phyllosilicates and oxyhydroxides, and (iii) to explore ion-exchange type rare earth element deposits for rare earth element procurement.

Nation	Production	Reserve
	metric tons	metric tons
Australia	18,000	5,700,000
Burma	38,000	not available
China	240,000	44,000,000
India	2900	6,900,000
Russia	2600	10,000,000
Thailand	7100	not available
United States	43,000	1,800,000
Brazil		21,000,000
Canada		14,000,000
Vietnam		22,000,000

Source: [5].

Table 1.
Rare earth oxide mine production and indicated reserves (2023).

2. Materials and methods

An aqua regia digestion was employed to obtain elemental abundances. Powdered samples (0.75 g) were equilibrated with 0.01 liter of aqua-regia (3 mole nitric acid:1 mole hydrochloric acid) in a 35°C incubator for 24 hours. Samples were shaken, centrifuged, and 0.45 μm filtered, with analysis using inductively coupled plasma emission—mass spectrometry (ICP-MS). A hot water extraction involved equilibrating 0.5 g samples in 0.02 L distilled-deionized water at 80°C for 1 hour followed by 0.45 μm filtering and elemental determination using ICP-MS. For the aqua regia digestion and water extraction, selected samples were duplicated, and reference materials were similarly analyzed to assess analytical accuracy. The Sharkey, Portageville, and Kaintuck soil series were collected in Dunklin, New Madrid, and Jefferson Counties, Missouri.

3. Representative rare earth element soil distributions

The Kaintuck soil series (coarse-loamy, siliceous, superactive, nonacid, mesic Typic Udifluvents) consists of very deep, well-drained soils formed in loamy alluvium. The Kaintuck soil aqua regia digestion concentrations of the rare earth elements are presented in **Figure 4**, and the corresponding rare earth element water extraction concentrations are presented in **Figure 5**. The Sharkey soil series (very fine, smectitic, thermic Chromic Epiaquerts) consists of very deep, poorly, and very poorly drained, very slowly permeable soils in clayey alluvium. The Sharkey soil aqua regia digestion concentrations of the rare earth elements are presented in **Figure 6**, and the corresponding rare earth element water extraction concentrations are presented in **Figure 7**. In both soil series, the aqua regia digestion has cerium exhibiting the greatest concentration, followed by Nd and La. The light rare earth elements concentrations are substantially greater than the heavy rare earth elements concentrations. The Kaintuck soil series shows Nd, Y, La, and Ce having the greatest water

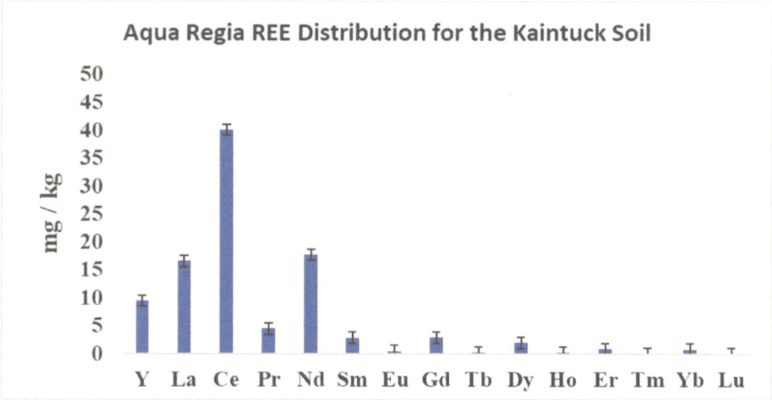

Figure 4.
The aqua regia digestion concentrations of the rare earth elements of the Kaintuck soil series. Error bars are the standard deviation for 20 observations.

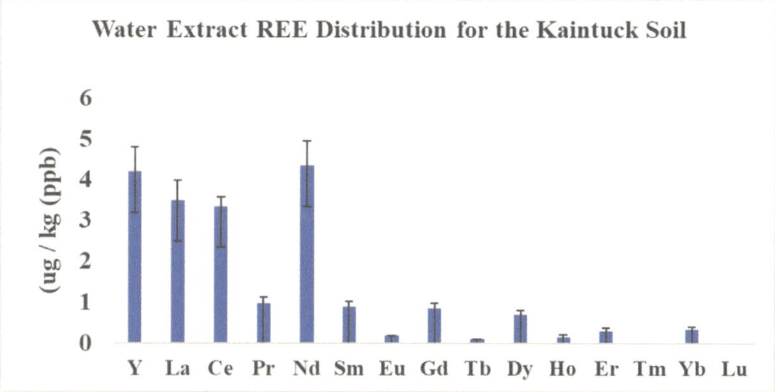

Figure 5.
The water extract concentrations of the rare earth elements of the Kaintuck soil series. Error bars are the standard deviation for 20 observations.

extraction concentrations, whereas the Sharkey soil series exhibits cerium having the greatest water extract recovery. The Kaintuck soil series has a mean cerium concentration of 40 mg Ce kg^{-1} (40,000 μg Ce kg^{-1}), whereas the corresponding cerium water extract concentration is approximately 3.2 μg Ce kg^{-1}. The Sharkey series presents similar rare earth element aqua regia digestion-water extraction concentration relationships. The large concentration differences between the aqua regia digestion and water extractable rare earth elements infer that a very substantial portion of the rare earth elements are less labile than the abundances associated with the water extraction pool.

4. Rare earth element hydrolysis with emphasis on cerium

Rare earth elements, typically exhibiting trivalent positive charges, undergo hydrolysis to reduce its charge polarization potential. Cerium hydrolysis for Ce^{3+} and

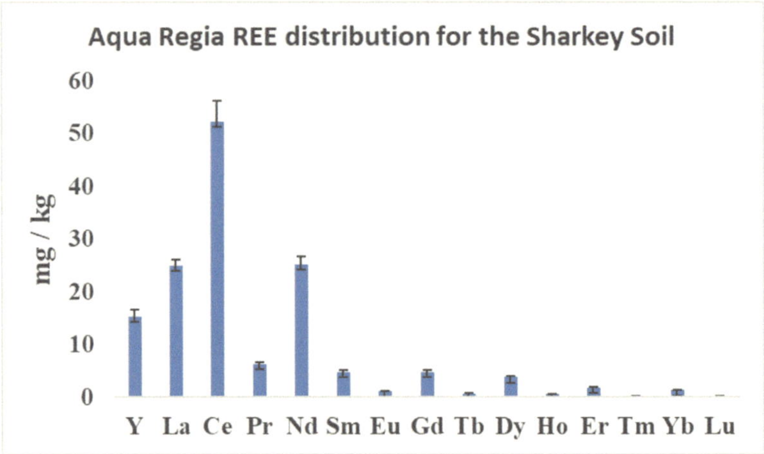

Figure 6.
The aqua regia digestion concentrations of the rare earth elements of the Sharkey soil series. Error bars are the standard deviation for seven observations.

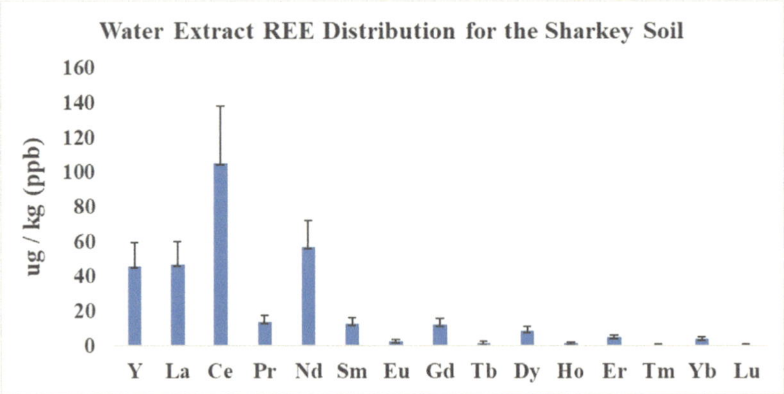

Figure 7.
The water extract concentrations of the rare earth elements of the Sharkey soil series. Error bars are the standard deviation for seven observations.

Ce^{4+} are presented (**Figures 8** and **9**). The hydrolysis constants for Ce^{3+} are displayed in Eqs. 1, 2, and 3.

$$[Ce(OH)][H+] = K_1[Ce][H_2O] \qquad K_1 = 10^{-8.41} \tag{1}$$

$$[Ce(OH)_2][H+]^2 = K_2[Ce][H2O]^2 \qquad K_2 = 10^{-17.6} \tag{2}$$

$$[Ce(OH)_3][H+]^3 = K_3[Ce][[H2O]^3 \qquad K_3 = 10^{-27.23} \tag{3}$$

Ce_T is the total analytical Ce aqueous concentration and was set at 0.001 molal. The pH intervals ranged from 4 to 9 in 0.2 pH units. The equilibrium solution's low

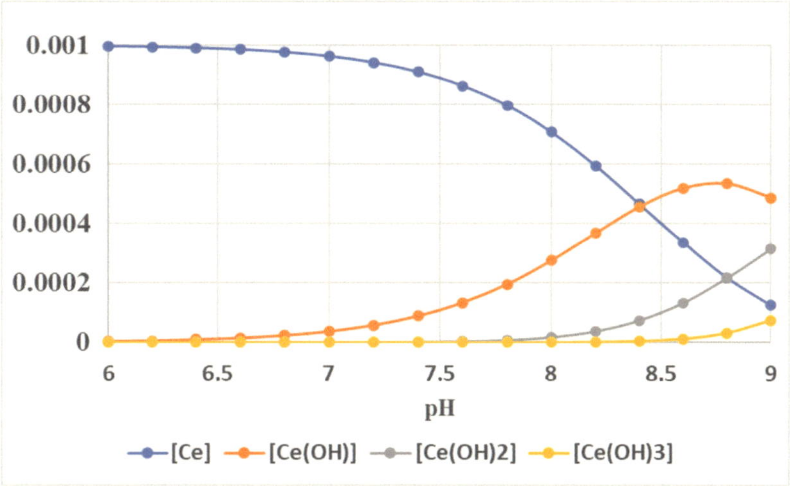

Figure 8.
Concentration distribution of Ce^{3+} hydrolysis species from pH 6 to 9.

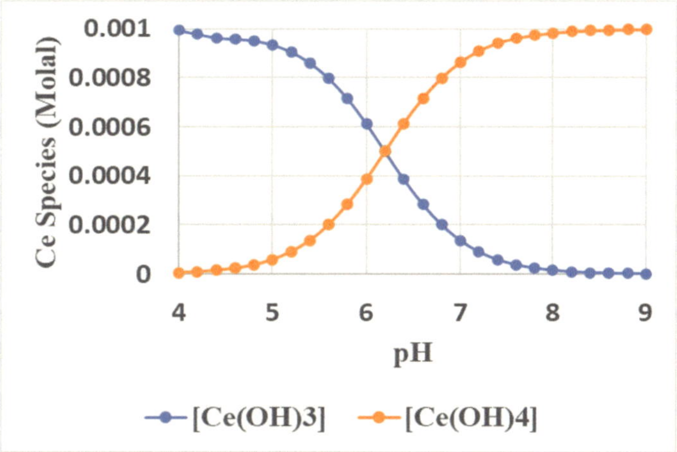

Figure 9.
Concentration distribution of Ce^{4+} hydrolysis species from pH 4 to 9.

ionic strengths yielded activity coefficients that were approximately 1, and thus activities and concentrations were essentially equal. The activity of water was 1. Equation 4 is the mass balance for Ce species.

$$Ce_T = [Ce] + [Ce(OH)] + [Ce(OH)_2] + [Ce(OH)_3] \tag{4}$$

Equation 5 was derived after the substitution of Eqs. 1, 2, and 3 into Eq. 4.

$$Ce_T = [Ce]\{1 + K_1/[H+] + K_2/[H+]^2 + K_3/[H+]^3\} \tag{5}$$

The hydrolysis species distribution of Ce^{3+} {[Ce], [Ce(OH)], [Ce(OH)$_2$], and [Ce(OH)$_3$]} is presented in **Figure 8**, showing that Ce^{3+} is the dominant Ce species from less than pH 6 to approximately pH 8.2.

For the hydrolysis of Ce^{4+}, the hydrolysis constants were $K_3 = 10^{-2.3}$ and $K_4 = 10^{-8.5}$, where the constants K_1 and K_2 are only relevant at pH levels considerably below pH 4 and typically outside of the natural pH interval of soils. Relevant rare earth element hydrolysis thermodynamic data have been tabulated [6–9]. Bentouhami et al. [9] provided log ß values for the following hydroxo Ce^{3+} species: $Ce(OH)^{2+}$ is −6.9, $Ce(OH)_3$ is −19.82, $Ce(OH)_4^{1-}$ is −28.9, $Ce_2(OH)_3^{3+}$ is −15.9, and $Ce_2(OH)_7^{1-}$ is −44.7. **Figure 9** illustrates that [$Ce(OH)_3^{+}$] is dominant from pH 5 to pH 6.2, whereas [$Ce(OH)_4$] is the dominant species in neutral and alkaline regimes.

5. Rare earth element complexation

Rare earth element complexation involving anionic species may be expressed as follows:

$$REE^{3+} + yL^{n-} = REELy^{(3-yn)}, \tag{6}$$

where L^{n-} is an inorganic ligand with a negative n charge and y is the stoichiometric coefficient. Common rare earth element anionic inorganic complexing species include nitrate, chloride, fluoride, sulfate, carbonate, and phosphate. Carbonate complexes ($REE(CO_3)^{+}$) are more prevalent in the light rare earth elements, and dicarbonate complexes ($REE(CO_3)_2^{-}$) are more prevalent in the heavy rare earth elements [10–12]. Luo and Byrne [13] documented the carbonate complexing behavior of the rare earth elements. For the Lanthanide Series, the dicarbonate complex becomes increasingly more stable with increasing atomic number. Millero [11] and Gramaccioli et al. [14] observed that rare earth elements-fluoride complexes obtained greater stability on transition from La to Lu. Banfield and Eggleton [15] emphasized rare earth elements-phosphate precipitates, especially apatites, in reducing soil rare earth element mobility.

Low molecular weight organic acid complexes with the rare earth elements include oxalic acid, citric acid, acetic acid, lactic acid, succinic acid, tartaric acid, formic acid, malic acid, and malonic acid, as well as naturally occurring higher molecular weight fulvic and humic acids [16–19]. As with the inorganic rare earth elements complexes, organic rare earth elements complexes tend to show greater stability for the heavy rare earth elements than the light rare earth elements [10, 17].

6. Cation exchange and adsorption reactions

Phyllosilicate-rare earth elements interactions have been examined to estimate the presence of lithologic discontinuities in soil and rare earth element mobility on soil weathering [4]. Nd and Eu adsorption on K-illite reflected outer-sphere exchange reactions and pH-dependent silanol and aluminol reactions [20–22]. Europium adsorption on Na- and Ca-montmorillonite was consistent with cation exchange reactions and strong and weak pH-dependent complexation reactions

[23–25]. These authors noted the presence of pH-dependent inner-sphere and outer-sphere adsorption sites. Tertre et al. [26–28] documented that REE^{3+} exchange involving montmorillonite and kaolinite was consistent with non-specific and specific complexation sites.

Coppin et al. [29] investigated rare earth element adsorption on Na-montmorillonite and kaolinite. Rare earth element adsorption was independent of electrolyte and dissolved CO_2 content; however, adsorption was dependent on clay mineralogy, pH, and ionic strength. Rare earth element adsorption increased at pH 5.5. The distribution coefficient (Kd) on montmorillonite declined slightly after pH 6. Stumpf et al. [30] observed Eu^{3+} adsorption onto smectite and kaolinite. At pH levels associated with soil environments inner-sphere Eu^{3+} surface complexes involving clay/Eu/carbonate were observed. Using regolith-hosted ion-adsorption deposits, Borst et al. [31] used synchrotron X-ray adsorption spectroscopy to indicate that REE‘s were adsorbed as outer-sphere hydrated complexes onto kaolinite. Zhenghua et al. [32] established La^{3+} adsorption isotherms on quartz, feldspar, kaolinite, goethite, and humic acid. The adsorption capacity increased in the order of humic acid > goethite ≈ kaolinite > feldspar ≈ quartz. More alkaline pH levels supported humic acid and goethite adsorption.

Investigating the Carrizo coarse-textured aquifer, Tang and Johannesson [33] documented the influence of carbonate ions and humic material activities on rare earth element adsorption as a function of CO_2, pH, ionic strength, and the initial rare earth element concentrations. They demonstrated that increased carbonate concentrations and increasingly alkaline pH levels decreased rare earth element adsorption. Subsequently, Tang and Johannesson [34] assessed the ability of soil organic matter to extract rare earth elements from coarse-textured coastal plain aquifer sediments. The recovery of rare earth elements was predicated on the ligand's concentration and complexation capability. Rare earth element surface stability constants from Fe-oxyhydroxides dominate the equilibria for rare earth element adsorption behavior if the ligand activities are small; however, if the ligand concentration is greater, the rare earth element fractionation patterns reflect the ligand activity. Shan et al. [35] demonstrated that the soil adsorption of La, Ce, Pr, and Nd was influenced by citric, malic, tartaric, and acetic acids, with adsorption capacity correlated with increasing cation exchange capacities and with increasing pH. In general, citric, malic, and tartaric acids suppressed rare earth element adsorption when compared to acetic acid or $Ca(NO_3)_2$.

Selecting eight soil profiles across Germany, Mihajlovic et al. [36] determined aqua regia digestion, hydroxylammonium chloride, and hot water extractable rare earth element concentrations to assess the potential for rare earth element mobilization. Silty and clayey textured soils showed greater rare earth element abundances than sandy and peaty textured soils. The hydroxylammonium chloride-extractable rare earth element concentrations inferred that Al, Fe, and Mn oxides and clay minerals were largely responsible for rare earth element adsorption. The hot water extractable rare earth element contents were greater in soils enriched in soil organic materials.

In selected alluvial soils, Aide [37] demonstrated Ce preferential accumulation on pedogenic Fe-Mn glaebules in soils having alternating oxidation-reduction conditions. In Taiwan, Wu et al. [38] examined rare earth elements to infer their potential mobility upon dissolution of pedogenic Fe oxides. Rare earth element abundances were positively associated with increased soil free iron oxide contents. Compared to the bulk soil, Fe oxyhydroxides exhibited preferences for heavy rare earth elements over the light rare earth elements. In mangrove soil environments, Andrade et al. [39] documented rare earth element abundances were in the decreasing order of the light rare earth elements, medium rare earth elements, and heavy rare earth elements. Fractionation of the light rare earth elements over the heavy rare earth elements was greater when kaolinite was the dominant phyllosilicate, and preferential heavy rare earth element adsorption was associated with Fe-rich smectite environments.

7. Ion adsorption-type rare earth element deposits

Ion adsorption-type rare earth element deposits are global rare earth element sources composed primarily of heavy rare earth elements and yttrium and are typically formed in weathered granitic regolith overlying granitic rocks [31, 39–41]. These deposits are also termed regolith-hosted ion-adsorption deposits and laterite-hosted rare earth deposits. In China, Kang et al. [39] observed regolith-hosted ion adsorption deposits formed by subtropical weathering of igneous rocks and demonstrated that electrostatically held rare earth elements were preferentially associated with kaolinite-halloysite regoliths [30].

In a recent and extensive review, Wu et al. [40] provided a general overview of the weathering sequence that culminates with the formation of rare earth element ion-adsorption deposits. Primary minerals are frequently associated with felsic igneous rocks, in subtropical climates weather, and are transformed to phyllosilicates. The initially formed secondary phyllosilicate illite is subsequently transformed to montmorillonite and, eventually with more intense weathering, to kaolinite and halloysite. With soil evolution, phyllosilicate eluviation-illuviation fosters clay textural differences, with rare earth element horizon concentration differences partially attributed to co-eluviation. Weathering further increases the degradation of rare earth-bearing minerals, particularly the more readily weatherable minerals (titanite, allanite, gadolinite, bastnasite, apatite, and others). Apatite weathering may release appreciable Ce^{3+}, which may be oxidized and recrystallized as cerianite (CeO_2). More weathering resistant minerals (zircon, monazite, and others) may remain as crystalline lattice structures. In southern Jiangxi Province, China, Li et al. [42] surveyed soils for their rare earth element contents and documented the following: (i) soil rare earth element abundances show spatial variation, (ii) concentrations of light rare earth elements are greater than the heavy rare earth elements concentrations, (iii) negative Ce anomalies and negative Eu anomalies were ubiquitous, and (iv) the soil rare earth element contents are primarily influenced by soil parent material inheritance.

Xiao et al. [43] focused on the adsorption thermodynamics of REEs on kaolinite to demonstrate that La, Nd, and Y adsorption conformed to Langmuir isotherms, and their saturated adsorption capacities were 1.73, 1.59 and 0.97 mg/g, respectively. Zhou et al. [44] investigated the sorption and desorption of Eu^{3+} on halloysite and kaolinite under slightly acidic conditions, noting that (i) at low NH_4^+ concentrations, halloysite adsorption of Eu is less than kaolinite, (ii) halloysite has a stronger capacity for REE enrichment than kaolinite, and (iii) Eu adsorption is primarily inner-sphere.

In ion-adsorption type deposits, rare earth element leaching is frequently facilitated by humic material complexation and formation of carbonate and phosphate ion-pairs. The heavy rare earth elements form stronger complex associations; therefore, these species may leach more readily and deeper into the regolith, permitting a greater degree of rare earth element fractionation [31–34]. Hydrolysis supports rare earth element adsorption as an exchangeable cation on phyllosilicates, again permitting the heavy rare earth elements to migrate into deeper regolith. The deeper regolith typically is less acidic, which supports the formation of more negatively charged sites along the phyllosilicate (primarily kaolinite) periphery, which may be represented as:

$$\text{-SOH} + \text{OH}^- \rightarrow \text{SO}^- + \text{H}_2\text{O} \tag{7}$$

$$\text{-SO}^- + \text{REE}^{3+} \rightarrow \text{-[SO - REE]}^{2+}, \tag{8}$$

where -[SO-REE]^{2+} represents REE complexation with aluminol and related functional groups.

In the Southeastern United States, Foley et al. [45] observed weathered granite-derived regoliths that have rare earth elements. The rare earth element chondrite-normalized patterns show tetrad patterns that infer their magmatic and weathering histories. Sanematsu and Watanabe [46] observed that rare earth element weathering associated with acidic soil environments will exhibit downward rare earth element transport. The rare earth elements, complexed with humic substances and carbonate-bicarbonates, are subsequently adsorbed or precipitated as secondary minerals in these less acidic environments. The secondary minerals include halloysite-kaolinite and phosphates. Cerium was adsorbed onto Mn oxides or oxidized-precipitated as cerianite (CeO_2). Fenga et al. [47] noted that the rare earth elements were preferentially surface adsorbed on kaolinite, and that adsorption was pH dependent with the maximum adsorption around pH 10. The rare earth element speciation was determined to be $REE(OH)_2^+$.

Yu et al. [48] investigated the granitoid bedrock of the Swedish Baltic Shield and focused on secondary mineral precipitates collected from fracture walls and groundwater samples. The data demonstrated that Ce(VI) was preferentially associated with Mn oxides in the upper portion of the fracture network. The authors inferred that repeated oxidation-reduction episodes were responsible for capturing Ce(VI). In a batch adsorption experiment in Madagascar, Berger et al. [49] investigated an intensely weathered soil profile and documented that the primary rare earth element minerals in the A horizon weather and illuviate and reconstitute as secondary rare earth element minerals in the developing B horizon. Cerium(IV) is associated with redox-sensitive elements (Mn, Fe) in B-horizon, and cerianite is preferentially accumulated. Li et al. [50] studied the Zudong deposit with heavy rare earth element adsorption onto halloysite and kaolinite in the deeper regolith. In rare earth element ion deposits, Yang et al. [51] noted that kaolinite and halloysite were the dominant phyllosilicates for rare earth element adsorption. The heavy rare earth elements were preferentially adsorbed at higher pH levels and greater ionic strengths.

In south China, Bao and Zhao [52] investigated rare earth element mineralization involving granite and acidic volcanic rocks, noting that granitic rock rare earth element mineralization is primarily controlled by the weathering resistance of rare earth accessory minerals. Well-developed weathering profiles typically showed three horizons: (i) a lateritic horizon, (ii) a weathered horizon, and (iii) the weathering front. Continuous leaching results in rare earth element accumulation, assuming that the erosion rate is appropriately limited. Exchangeable rare earth elements were recoverable with complexing agents. Cerium was enriched in the upper most layer, a feature attributed to Ce oxidation and cerianite formation or excessive clay and Fe and Al oxyhydroxide adsorption.

The surface mining and heap leaching of China's unique ion-adsorption rare earth resources have caused severe environmental damage [53]. Liu et al. [54] investigated ion-adsorption rare earth mine sites and documented excessive concentrations of rare earth element in soils, streams, and sediments. Enriched middle rare earth element concentrations suggested leaching and Al and Fe oxyhydroxide adsorption, coupled with ligand complexation, were important processes in rare earth element fractionation. Moldoveanu and Papangelakis [55] determined that (i) rare earth elements adsorbed on phyllosilicates are recovered by leaching with simple salt solutions, (ii) the leaching efficiency decreased in the order $Cs^+ > NH_4^+ > Na^+ > Li^+$, and (iii) sulfate anions confer greater recoveries than chlorides.

8. Summary

Ion-adsorption rare earth element deposits are an area of interest given the increasing utility of rare earth elements in advanced technologies. The framework for understanding these potentially lucrative rare earth element resources is rapidly evolving. Our existing understanding of rock weathering, soil formation, and soil processes is appropriately being applied to support the evaluation of these resources for rare earth element provision and the development of viable and sustainable extraction and processing.

9. Further research aims critical to rare earth element supply

Technology procurement and intensification is predicated on the acquisition of rare earth elements, especially cerium. In accordance, future rare earth element research must focus on (i) furthering our understanding of REE relationships in soil, including hydrolysis, complexation, and adsorption and (ii) developing environmentally sustainable mining and processing protocols involving ion-adsorption rare earth element deposits. A more sophisticated understanding of ion-adsorption REE deposit formation supports future improvements in mining site classification and the provision of economically advantageous recovery and processing technologies.

Conflict of interest

The author has no conflict of interest.

Author details

Michael Aide
Southeast Missouri State University, Cape Girardeau, MO, USA

*Address all correspondence to: mtaide@semo.edu

References

[1] Lee JD. Concise Inorganic Chemistry. London, UK: Chapman Hall; 1991

[2] Pendias K. Trace Elements in Soils and Plants. NY: CRC Press; 2011

[3] Aide MT, Aide CC. Rare earth elements: Their importance in understanding soil genesis. International Scholarly Research Network (ISRN Soil Science). 2012;**2012**. Article ID 783876. DOI: 10.5402/2012/783876

[4] Aide MT. Lanthanide soil chemistry and its importance in understanding soil pathways: Mobility, plant uptake and soil health. In: Lanthanides. London, UK, Rijeka, Croatia: InTech; 2018. DOI: 10.5772/Intechopen.79238

[5] Van Veen K, Melton A. Rare Earth Elements Supply Chains, Part 1: An Update on Global Production and Trade. United States International Trade Commission Executive Briefing on Trade; 2020. Available from: https://www.usitc.gov/publications/332/executive_briefings/ebot_rare_earths_part_1.pdf

[6] Baes CF, Messmer RE. The Hydrolysis of Cations. New York: Wiley-Interscience Publication; 1976

[7] Hummel E, Berner U, Curti E, Thoenen A. Nagra/PSI Chemical Thermodynamic Data Base. Wettingen, Switzerland: Nagra; 2008

[8] Klungness GD, Byrne RH. Comparative hydrolysis behavior of the rare earths and yttrium: The influence of temperature and ionic strength. Polyhedron. 2000;**19**:99-107. DOI: 10.1016/S0277-5387(99)00332-0

[9] Bentouhami E, Bouet GM, Meullemeestre J, Vierling F, Khan MA. Physicochemical study of the hydrolysis of rare-earth elements (III) and thorium (IV). Comptes Rendus Chimie. 2004;**7**:537-545. DOI: 10.1016/j.crci.2004.01.008

[10] Cantrell KJ, Byrne RH. Rare earth element complexation by carbonate and oxalate ions. Geochimica et Cosmochimica Acta. 1987;**51**:597-605

[11] Millero FJ. Stability constants for the formation of rare earth-inorganic complexes as a function of ionic strength. Geochimica et Cosmochimica Acta. 1992;**56**:3123-3132

[12] Lee JH, Byrne JH. Complexation of trivalent rare earth elements (Ce, Eu, Gd, Tb, Yb) by carbonate ions. Geochimica et Cosmochimica Acta. 1993, 1993;**57**(2):295-302

[13] Luo YR, Byrne JH. Carbonate complexation of yttrium and the rare earth elements in natural waters. Geochimica et Cosmochimica Acta. 2000;**68**:691-699

[14] Gramaccioli CM, Diella V, Demartin F. The role of fluoride complexes in REE geochemistry and the importance of 4f electrons: Some examples in minerals. European Journal of Mineralogy. 1999;**11**:983-992

[15] Banfield JF, Eggleton RA. Apatite replacement and rare earth mobilization, fractionation, and fixation during weathering. Clays Clay Minerals. 1989;**37**:113-127

[16] Gu Z, Wang X, Gu X, Cheng J, Wang L, Dai L, et al. Determination of stability constants for rare earth elements and fulvic acids extracted from different soils. Talanta. 2001;**53**:1163-1170. DOI: 10.1016/s0039-9140(00)00606-8

[17] Schijf J, Byrne RH. Stability constants for mono-and dioxalato-complexes of Y and the REE, potentially important species in groundwaters and surface freshwaters. Geochimica et Cosmochimica Acta. 2001;**65**:1037-1046

[18] Marsac R, Catrouillet C, Davranche M, Bouhnik-Le Coz M, Briant N, Janot N, et al. Modeling rare earth elements binding to humic acids with model VII. Chemical Geology. 2021;**567**. Article ID 120099. DOI: 10.1016/j.chemgeo.2021.120099

[19] Pourret O, Davranche M, Gruau G, Dia A. Competition between humic acid and carbonates for rare earth elements complexation. Journal of Colloid and Interface Science. 2007;**305**:25-31

[20] Sinitsyn VA, Aja SU, Kulik DA, Wood SA. Acid-base surface chemistry and sorption of some lanthanides on K+-saturated, Marblehead illite: I. Results of an experimental investigation. Geochimica et Cosmochimica Acta. 2000;**64**:185-194

[21] Kulik DA, Aja SU, Sinitsyn VA, Wood SA. Acid-base surface chemistry and sorption of some lanthanides on K+, Marblehead illite: II. A multisite-surface complexation modeling. Geochimica et Cosmochimica Acta. 2000;**64**:195-213

[22] Rabung T, Pierret MC, Bauer A, Geckeis H, Bradbury MH, Baeyens B. Sorption of Eu(III)/Cm(III) on Ca-montmorillonite and Na-illite. Part 1: Batch sorption and time-resolved laser fluorescence spectroscopy experiments. Geochimica et Cosmochimica Acta. 2005;**69**:5393-5402

[23] Bradbury MH, Baeyens B, Geckeis H, Rabung T. Sorption of Eu(III)/Cm(III) on Ca-montmorillonite and Na-illite. Part 2: Surface complexation modelling. Geochimica et Cosmochimica Acta. 2005;**69**:5403-5412

[24] Bradbury MH, Baeyens B. Sorption of Eu on Na- and Ca-montmorillonites: Experimental investigations and modelling with cation exchange and surface complexation. Geochimica et Cosmochimica Acta. 2002;**66**:2325-2334

[25] Tertre EG, Berger S, Castet M, Loubet GE. Experimental sorption of Ni. Cs and ln onto a montmorillonite up to 150°C. Geochimica et Cosmochimica Acta. 2005;**69**:3937-4948

[26] Tertre EG, Berger E, Giffaut M, Loubet CH. Europium retention onto clay minerals from 25 to 150°C: Experimental measurements, spectroscopic features and sorption modeling. Geochimica et Cosmochimica Acta. 2006;**70**:4563-4578

[27] Tertre E, Hofmann A, Berger C. Rare earth element sorption by basaltic rock: Experimental data and modeling results using the "generalised composite approach". Geochimica et Cosmochimica Acta. 2008;**72**:1043-1056

[28] Coppin F, Berger G, Bauer A, Castet S, Loubet M. Sorption of lanthanides on smectite and kaolinite. Chemical Geology. 2002;**182**:57-68. DOI: 10.1016/S0009-2541(01)00283-2

[29] Stumpf T, Bauer A, Coppin F, Fanghänel T, Kim JI. Inner-sphere, outer-sphere and ternary surface complexes: A TRLFS study of the sorption process of Eu(III) onto smectite and kaolinite. Radiochimica Acta. 2002;**90**:345-349. DOI: 10.1524/ract.2002.90.6.345

[30] Borst AM, Smith MP, Finch AA, Estrade G, Villanova-de-Benavent C, Nason P, et al. Adsorption of rare earth elements in regolith-hosted clay deposits. Nature Communications. 2020;**11**:4386. DOI: 10.1038/s41467-020-17801-5

[31] Zhenghua W, Jun L, Hongyan G, Xiaorong W, Chunsheng Y. Adsorption

isotherms of lanthanum to soil constituents and effects of pH, EDTA and fulvic acid on adsorption of lanthanum onto goethite and humic acid. Chemical Speciation and Bioavailability. 2001;**13**:75-81. DOI: 10.3184/095422901782775444

[32] Tang J, Johannesson KH (2010) Rare earth elements adsorption onto Carrizo sand: Influence of strong solution complexation. Chemical Geology 279:120-133. https://doi.org/10.1016/j.chemgeo.2010.10.011

[33] Tang J, Johannesson KH. Ligand extraction of rare earth elements from aquifer sediments: Implications for rare earth element complexation with organic matter in natural waters. Geochimica et Cosmochimica Acta. 2010;**74**:6690-6705. DOI: 10.1016/j.gca.2010.08.028

[34] Shan XQ, Lian J, Wen B. Effect of organic acids on adsorption and desorption of rare earth elements. Chemosphere. 2002;**47**:701-710. DOI: 10.1016/s0045-6535(02)00032-2

[35] Mihajlovic J, Bauriegel A, Stärk HJ, Roßkopf N, Zeitz J, Milbert G, et al. Rare earth elements in soil profiles of various ecosystems across Germany. Applied Geochemistry. 2019;**102**:197-217. DOI: 10.1016/j.apgeochem.2019.02.002

[36] Aide MT. Elemental composition of soil nodules from two Alfisols on an alluvial terrace in Missouri. Soil Science. 2005;**170**:1022-1033

[37] Wu CY, Chu MF, Huang KF, Hseu ZY. Rare earth elements associated with pedogenic iron oxides in humid and tropical soils from different parent materials. Geoderma. 2022;**423**:115966. DOI: 10.1016/j.geoderma.2022.115966

[38] Andrade GRP, Cuadros J, Barbosa JMP, Vidal-Torrado P. Clay minerals control rare earth elements (REE) fractionation in Brazilian mangrove soils. Catena. 2022;**209**:105855. DOI: 10.1016/j.catena.2021.105855

[39] Kang X, Zhao Z, Yang X, Liu Q. High light REE enrichments in weathering crust in the Huangmeijian alkaline pluton in Luzong Volcanic Basin: A potential exploration target for ion-adsorption rare earth element deposit. Solid Earth Sciences. 2023;**8**(1):29-33. DOI: 10.1016/j.sesci.2023.01.001

[40] Wu Z, Chen Y, Wang Y, Xu Y, Lin Z, Liang X, et al. Review of rare earth element (REE) adsorption on and desorption from clay minerals: Application to formation and mining of ion-adsorption REE deposits. Ore Geology Reviews. 2023;**157**. Article ID 105446. DOI: 10.1016/j.oregeorev.2023.105446

[41] Chen H, Chen L, Zhang L, Guo M. Spatial heterogeneity of rare earth elements: Implications for the topsoil of regional ion-adsorption type rare earth deposit areas in southern China. Minerals. 2023;**13**(6):784. DOI: 10.3390/min13060784

[42] Li W, Zuo Y, Wang L, Wan X, Yang J, Liang T, et al. Abundance, spatial variation, and sources of rare earth elements in soils around ion-adsorbed rare earth mining areas. Environmental Pollution. 2022;**15**(313):120099. DOI: 10.1016/j.envpol.2022.120099

[43] Xiao Y, Huang L, Long Z, Feng Z. Adsorption ability of rare earth elements on clay minerals and its practical performance. Journal of Rare Earths. 2016;**34**(5):543-548. DOI: 10.1016/S1002-0721(16)60060-1

[44] Zhou J, Liu H, Liu D, Yuan P, Bu H, Du P, et al. Sorption/desorption of Eu(III) on halloysite and kaolinite.

Applied Clay Science. 2022;**216**:106356. DOI: 10.1016/j.clay.2021.106356

[45] Foley N, Ayuso R. REE enrichment in granite-derived regolith deposits of the southeastern United States: Prospective source rocks and accumulation processes. Geological Survey Paper. 2015;**3**:131-138

[46] Sanematsu K, Watanabe Y. Characteristics and genesis of ion adsorption-type rare earth element deposits. Reviews in Economic Geology. 2016;**18**:55-79. DOI: 10.5382/Rev.18.03

[47] Fenga X, Onelc O, Council-Trocheb M, Noblec A, Yoonc RH, Morrisa JR. A study of rare earth ion-adsorption clays: The speciation of rare earth elements on kaolinite at basic pH. Applied Clay Science. 2021;**201**:105920. DOI: 10.1016/j.clay.2020.105920

[48] Yu C, Drake H, Mathurin FA, Åström ME. Cerium sequestration and accumulation in fractured crystalline bedrock: The role of Mn-Fe (hydr-) oxides and clay minerals. Geochimica et Cosmochimica Acta. 2017;**199**:370-389. DOI: 10.1016/j.gca.2016.11.044

[49] Berger A, Janots E, Gnos E, Frei R, Bernier F. Rare earth element mineralogy and geochemistry in a laterite profile from Madagascar. Applied Geochemistry. 2014;**41**:218-228. DOI: 10.1016/j.apgeochem.2013.12.013

[50] Li MYH, Zhou MF. The role of clay minerals in forming the regolith-hosted heavy rare earth element deposits. American Mineralogist. 2019;**105**:92-108. DOI: 10.2138/am-2020-7061

[51] Yang M, Liang X, Ma L, Huang J, He H, Zhu J. Adsorption of REEs on kaolinite and halloysite: A link to the REE distribution on clays in the weathering crust of granite. Chemical Geology. 2019;**525**:210-217. DOI: 10.1016/j.chemgeo.2019.07.024

[52] Bao Z, Zhao Z. Geochemistry of mineralization with exchangeable REY in the weathering crusts of granitic rocks in South China. Ore Geology Reviews. 2008;**33**(3-4):519-535. DOI: 10.1016/j.oregeorev.2007.03.005

[53] Yang XJ, Lin A, Li XL, Wu Y, Zhou W, Chen Z. China's ion-adsorption rare earth resources, mining consequences and preservation. Environmental Development. 2013;**8**:131-136. DOI: 10.1016/j.envdev.2013.03.006

[54] Liu WS, Guo MN, Liu C, Yuan M, Chen XT, Huot H, et al. Water, sediment and agricultural soil contamination from an ion-adsorption rare earth mining area. Chemosphere. 2018;**216**:75-83. DOI: 10.1016/j.chemosphere.2018.10.109

[55] Moldoveanu GA, Papangelakis VG. Recovery of rare earth elements adsorbed on clay minerals: I. Desorption mechanism. Hydrometallurgy. 2012;**117-118**:71-78. DOI: 10.1016/j.hydromet.2012.02.007

Section 2

Emerging Technology Utilizing Cerium

Chapter 2

Cerium Compounds in LEDs Technology

Machaba L.A. Letswalo and Leelakhrishna Reddy

Abstract

This chapter explores the application and benefits of cerium compounds in LEDs, delving into their impact on photoluminescence, electroluminescence, quantum efficiency, and color quality. Cerium compounds (CeO_2) play a crucial role in enhancing the efficiency of LEDs. Phosphors are substances that absorb photons and emit light of different wavelengths. By introducing cerium compounds as phosphors in LEDs, the conversion of blue light to other colors becomes more efficient, resulting in improved luminous efficacy. In this chapter, the study on YAG:Ce phosphor material shows that vacuum annealing tends to suppress Ce^{3+} oxidation more than air annealing, leading to a considerable 20% increase in photoluminescence emission intensity. The optimized YAG:Ce nanoparticles were used to create a white light emitting diode (WLED) that produced cool-white light with a maximum radiation luminous efficacy of 285 lm/W and a comparable color rendering index of 83. Furthermore, $Na_2MgSiO_4:Ce^{3+}:Li^+$ phosphor study displays tunable emission color characteristics when the content of Ce^{3+} is increased or Li^+ is added as a charge compensator. The utilization of cerium compounds (CeO_2) has demonstrated significant efficiency and color quality improvements on YAG:Ce and $Na_2MgSiO_4:Ce^{3+}:Li^+$ phosphor components.

Keywords: CeO_2, photoluminescence, electroluminescence, quantum efficiency, LEDs

1. Introduction

Cerium compounds are a fascinating material class with unique optical and luminescent properties. These compounds contain the element cerium and have garnered significant interest in scientific research due to their wide range of applications in various fields. In this chapter, we will briefly overview cerium compounds and delve into the importance of their optical and luminescent properties.

1.1 Overview of cerium compounds

Cerium is a rare-earth metal with many industrial applications, making it valuable. Cerium compounds slowly react with water, and commercial-grade cerium has an iron-gray coloration and is as ductile as tin when in its pure form [1]. As the most abundant rare-earth metal, cerium oxidizes in air at room temperature to form CeO_2 [1]. Cerium compounds have industrial significance and are used in mantles of lanterns, ceramic, photographic, and textile industries, as well as in analytical chemistry [1]. Cerium oxide (CeO_2) have oxygen storage capacities, optical properties, and catalytic properties that make them highly useful, and they have been investigated as an oxygen scavenger in biological systems in drugs and pharmaceutical preparations [2]. CeO_2 NPs have various industrial applications, including use as a polishing agent, ultraviolet absorbing compound, solid electrolytes, fuel additive, luminescence and automotive exhaust catalyst [2]. Cerium exists in two main oxidation states, Ce (III) and Ce (IV), with the latter being the only lanthanide that has important aqueous and coordination chemistry in the +4-oxidation state [3, 4]. Ceria is used in chemical-mechanical planarization (CMP) for polishing [4]. However, cerium metal reacts slowly with water and quickly dissolves in diluted acids except hydrofluoric acid [5]. Cerium is a moderately strong paramagnet that becomes antiferromagnetic below 13 K and becomes superconducting in the millikelvin range at pressures exceeding 20 kbar [5]. Cerium is not toxic when eaten, animals injected with large doses of cerium have died due to cardiovascular collapse, and it is a rare-earth metal with low to moderate toxicity [4]. Cerium is a strong reducing agent and ignites spontaneously in air at 65 to 80°C, and fumes from cerium fires are toxic [4]. Ultimately, cerium has many industrial applications, including use in petroleum cracking catalysts, three-way automotive emission catalysts, and ferrous alloys used to scavenge sulfur and oxygen and to nodulize cast iron [5].

1.2 The significance of optical and luminescent properties of cerium compounds in scientific

Cerium compounds have been extensively studied in scientific research and various industrial applications due to their remarkable optical and luminescent properties. A study was conducted to investigate the optical and scintillation properties of cerium-doped lanthanide trihalides, aimed at gaining a better understanding of the scintillation process in this trihalide series [6]. The study reported emission and excitation spectra, X-ray induced emission spectra, pulse-height measurements, and scintillation decay time spectra. The fluorescence spectra exhibited structure at low temperatures, while garnet crystals showing luminescence exhibited bright-yellow fluorescence [6, 7]. Cerium-activated garnet crystals exhibited luminescence, while the wavelength maxima of the fluorescence-spectrum profiles were found to

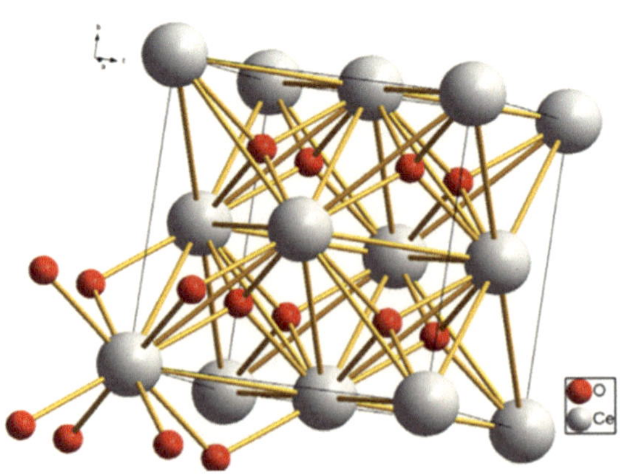

Figure 1.
Visualization of the crystal CeO_2 structure with the cubic space group of Fm-3 m (225), (ICDD – 43-1002).

be host-dependent [7]. Additionally, the direct bandgap of cerium oxide falls in the 3.2–3.6 eV range, while the indirect bandgap falls between 2.9–3.3 eV [8]. Cerium oxide have excellent optical properties, making them useful for electro-optical and optoelectronic devices [8]. Ce^{3+} ion is an important activator whose excitation and emission are based on the transition between its 4f ground levels and 5d excited levels. The energy of 5d levels of Ce^{3+} relative to 4f ground state depends on its coordination environment that relies on the host structure. According to this feature, broadband emission with different colors can be achieved through the adjustment of the host structure. Therefore, the activator generally has a similar ionic radius and charge to the substituted ion. The crystal cubic fluorite-type of space group Fm-3 m structure of the CeO_2 compound is presented in **Figure 1** [6–8].

2. Synthetization

Various methods have been developed to produce nanostructured materials from gas, liquid, or solid phases. In recent years, materials such as cerium compounds have been used to produce nanostructures as a host (base) or activator and we discuss some synthesis methods below [9, 10].

2.1 Combustion synthesis of nanocrystalline CeO_2-based powders via ethylene glycol: Nitrate process

In accordance with the specified ethylene glycol-nitrate ratios, precise quantities of cerium nitrate ($Ce(NO_3)_3{\bullet}6H_2O$ 99.5%) and ethylene glycol (99%) were meticulously blended in deionized water, ensuring the production of transparent solutions. These solutions were subsequently evaporated in an insulated tall beaker placed on a hot plate at a temperature of 105°C. Moreover, a proportion of the transparent solutions underwent additional drying in an oven at the same temperature for a duration of 24 hours, resulting in the formation of a gel precursor employed for TGA-DSC analysis. Once the viscous liquids took shape, a finely woven lid was securely positioned on the beaker to capture the ensuing powder during the ensuing autoignition process. Concurrently, the temperature of the hot plate was elevated to 250°C, leading to the expansion and autoignition of the viscous liquids, ultimately culminating in the rapid discharge of a substantial volume of gases and the eventual formation of voluminous powders. To obtain pure and well-crystallized ceria powders, the powders obtained post-autoignition were subjected to calcination at 500°C for a duration of 2 hours utilizing a muffle furnace. Subsequently, the calcined powders were milled for 6 hours in ethanol utilizing a planetary mill rotating at a rate of 350 r/min. They were subsequently dried and further pulverized in a mortar [9].

2.2 Synthesis of cerium(IV) oxide ultrafine particles by solid-state reactions

For the experimental procedure, a number of chemicals were utilized, specifically AR-grade ammonium cerium (IV) nitrate [(NH4)2Ce(NO3)6], cerium(III) nitrate hexahydrate [Ce(NO3)3z6H2O], sodium hydroxide (NaOH), and anhydrous ethanol (C2H5OH). To obtain ultrafine particles of CeO2, a mixture of (NH4)2Ce(NO3)6 (0.01 mol, 5.48 g) and NaOH (0.06 mol, 2.40 g) was initially combined and ground at room temperature for 30 minutes under normal air conditions. The resulting mixture was then suspended in distilled water and underwent 10 minutes of ultrasound treatment. Subsequently, the solid product was subjected to centrifugation for 10 minutes at 8000 rpm. This washing procedure was repeated using 80 mL of distilled water until the solution's pH dropped below 8. Afterwards, the product was washed twice with 40 mL of ethanol, dried in air, and yielded CeO2 ultrafine particles weighing 1.69 g, with a 98% yield. A portion of the obtained solid CeO2 (approximately 0.25 g) was then sintered at various temperatures ranging from room temperature to 800°C for a duration of 2 hours in the presence of air. Additionally, Ce(NO3)3z6H2O (0.01 mol, 4.34 g) and NaOH (0.03 mol, 1.20 g) underwent a similar reaction to produce CeO2 ultrafine particles [10].

Below are synthesis methods where cerium compounds are commonly used as activators.

2.3 Solid state method

The MgB_4O_7 powders were prepared by the solid-state synthesis method. MgO (Merck, 99.9% purity) and H_3BO_3 (Merck, 99.9% purity) were selected as starting reactants. The synthesis process began with mixing certain amounts of MgO and H_3BO_3 in an agate mortar and grinding until a homogeneous mixture was obtained. After the grinding operations, the mixture was calcined in an ash kiln at 850°C for 6 hours. The Ce dopant was selected in oxide form cerium oxide (CeO_2, Merck, powder, 99.995%) and added to the mixture in different weight ratios. The dopant-containing mixture was ground again and calcined the next time at 850°C for 6 hours [11, 12].

2.4 Solution combustion method

MgB_4O_7 was synthesized by solution combustion synthesis (SCS). This method combines a metal nitrate, fuel and other precursors in water and heats the mixture until combustion, producing a form of fine particles of the product. In the case of MBO, magnesium nitrate hexahydrate, boric acid, nitrates of the desired dopants (mainly $LiNO_3$ and $Ce(NO_3)_3$) and urea are mixed. An excess of boric acid is required to obtain the desired MgB4O7 crystal phase. In addition, the use of high-purity reagents to improve the material was investigated. After synthesis, annealing is required to crystallize the material fully: samples were heated to 900°C at 5°C/min and soaked for 2 h at the selected annealing temperature and let it cool down [13, 14].

3. Characterization

The optical absorption spectrum of phosphor materials was recorded using a Perkin Elmer UV-Vis spectrophotometer at room temperature. The photoluminescence (PL) and lifetime measurements were carried out in a spectro-fluorometer with a xenon lamp as an excitation source.

3.1 Optical properties of Ce based MgB_4O_7 compounds

Band gap energy, often simply called band gap, is a fundamental concept in solid state physics and materials science. It is a key property that characterizes the electronic structure of semiconductors and insulators. The band gap represents the energy difference between the valence band and the conduction band in the electronic band structure of the material. The size of the band gap determines the electrical properties of the material. Bandgap energy is a key property when selecting materials for specific applications in electronics, photonics and optoelectronics. Different materials have different bandgap energies, and engineers and scientists choose materials with an appropriate bandgap for their desired function. It is a basic property that characterizes materials' electrical conductivity and optical properties [15].

Materials with a band gap larger than the energy of visible light (typically 1.65 to 3.1 eV) are transparent to visible light. A material's color depends on the energy of the absorbed light, which corresponds to the energy of the band gap. This energy threshold has several important implications for a material's interaction with light, such as its transparency, color and ability to absorb or emit photons [16].

Bandgap energy analysis of cerium (Ce) compounds involves studying these materials' electronic structure and properties. Cerium is a rare earth element that can form a variety of compounds with different band gap energies, depending on the specific compound and its crystal structure. For example, analysis of the bandgap energy of cerium compounds involves a combination of theoretical calculations, experimental measurements, crystal structure analysis, and an understanding of the compound's intended application. The energy of the bandgap determines the wavelength of light that a semiconductor can emit or absorb. For example, in light-emitting diodes (LEDs), the bandgap energy determines the color of the light produced. Different materials with different bandgap energies can be used to make LEDs of different colors [16–19].

3.2 Excitation spectrum

An excitation spectrum is a graphical representation of how the absorption or excitation of a substance changes with the wavelength or energy of the incident light. In an excitation process, a substance is exposed to light and can absorb photons whose energies correspond to the energy levels or transitions within the substance. This absorption causes electrons to move into higher energy states or orbitals.

3.3 Emission spectrum

An emission spectrum is a graphical representation of the light emitted by a substance as it returns to lower energy states after excitation and shows the emitted photons. As the substance returns to its lower energy state, it releases energy in the form of photons. These emitted photons have energies corresponding to the energy levels or transitions involved in the de-excitation process.

3.4 Relationship

The emission lines in the emission spectrum correspond to the same energy levels or transitions observed in the excitation spectrum. This correspondence is a fundamental aspect of the Kasha-Vavilov rule, which states that the emission energy (wavelength) in a photoluminescence process should be independent of the excitation wavelength. It allows researchers to identify the specific energy levels and transitions that play a role in the luminescence behavior of cerium compounds [20].

The excitation and emission spectra of a material are influenced by various factors, including the material's electronic structure, chemical composition, physical state, and external conditions. The main key factors that can influence the excitation and emission spectra are chemical compositions, crystal structures, dopant/impurities, pressure, chemical environment temperature, concentration, excitation source, molecular structure, etc., [21].

These factors are essential in studying the excitation and Emission spectra of materials. Researchers often manipulate these factors to develop and optimize materials for various applications, including fluorescence, phosphorescence, laser operation, and optoelectronic devices [22].

4. Luminescent properties of cerium compounds

4.1 Impact of doping of cerium compound with other elements that influences its optical and luminescent properties

Doped cerium compounds have been intentionally mixed with other elements or molecules. For instance, by doping cerium compounds with specific rare earth elements, researchers can fine-tune the emission spectra of LEDs, expanding the range of available colors. Additionally, doped cerium compounds can enhance the stability and longevity of LEDs. By carefully selecting dopants, manufacturers can mitigate issues such as color shift and degradation of performance over time, ensuring the longevity and reliability of LED lighting systems.

4.2 Luminescent properties of various host materials doped with cerium compound

4.2.1 Definition of photoluminescence

Photoluminescence is defined as the emission of light from a material upon its absorption of photons. In the case of cerium compounds, this phenomenon occurs due to the excitation of electrons in the outermost orbitals of the cerium ions. Upon absorption of photons, these electrons gain energy and are promoted to higher energy levels, leaving behind holes in the lower energy levels. Subsequently, the recombination of electrons and holes leads to the emission of light. Below, the process of the photoluminescence mechanism and the analysis of the emission spectra of different host material doped cerium compounds as an activator.

4.2.2 Electroluminescence properties

Electroluminescence, on the other hand, is the emission of light from a material upon applying an electric field. Cerium compounds have demonstrated promising properties in electroluminescent devices. These compounds can be incorporated into phosphors, such as organic light-emitting diodes (OLEDs), to create efficient and customizable sources of light. The advantageous characteristics of cerium compounds in electroluminescence applications include their high luminous efficiency, broad emission spectra, and long operational lifetimes. These compounds can be easily tailored to emit light across a wide range of colors, making them suitable for various display applications. However, it is crucial to acknowledge the limitations of cerium compounds in electroluminescence applications. The use of cerium compounds as an activator in electroluminescent devices—where the cerium compound is doped to the host material—has been investigated below.

4.2.3 Quantum efficiency of cerium compound

Quantum efficiency is a measure of the efficiency with which a material converts absorbed photons into emitted photons. Evaluating the quantum efficiency of cerium compounds is crucial to assessing their performance relative to other luminescent materials. Several factors, including the purity of the compound, defects or impurities present, and the electronic structure of the host lattice, influence the quantum efficiency of cerium compounds.

Comparative studies below have revealed that cerium compounds exhibit competitive quantum efficiency when compared to other luminescent materials. Their unique energy levels and electron configurations allow for excellent photon conversion efficiency. This characteristic, coupled with their advantageous photoluminescence and electroluminescence properties, positions cerium compounds as promising candidates for various applications in the field of luminescent materials.

5. Photoluminescence properties of YAG are doped with a cerium compound (Ce^{3+} ion)

The PL emission spectra of the annealed and as-combusted YAG:Ce NPs are displayed in **Figure 2** at 224 nm. Without an ordered crystalline structure, the activator was unable to integrate into the crystal lattice, which prevented the formation of a luminescent center necessary for optical transition. The d-f emission was typically composed of double structures that made up the characteristic long wavelength emission because of the strong crystal field applied to the dodecahedral site in the YAG structure where the Ce^{3+} ions are located. In the emission band, a double Gaussian function with an energy difference of roughly 1923 cm^{-1} was deconvoluted from the $^5d_1 \rightarrow {}^2F_{(5/2)}$ (525 nm) and $^5d_1 \rightarrow {}^2F_{(7/2)}$ (577 nm) transitions of the Ce^{3+} ion, as shown in **Figure 2(a)** insert. The schematic illustration presented in **Figure 2b** provides insight into the optical transition mechanism for the Ce^{3+}ion in garnet crystal structure [24, 25].

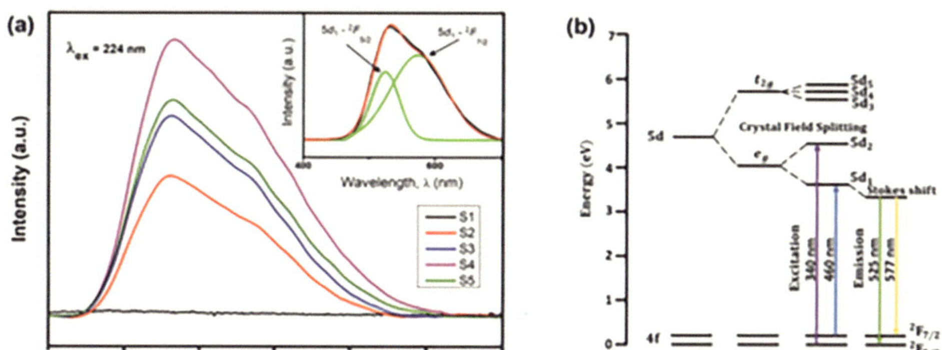

Figure 2.
PL emission spectra for the as-combusted and different annealing treated YAG:Ce NPs (a) and schematic diagram of the electronic level structure of Ce^{3+} in YAG lattice [23]. The inset of (a) shows the deconvoluted emission band for YAG:Ce NPs.

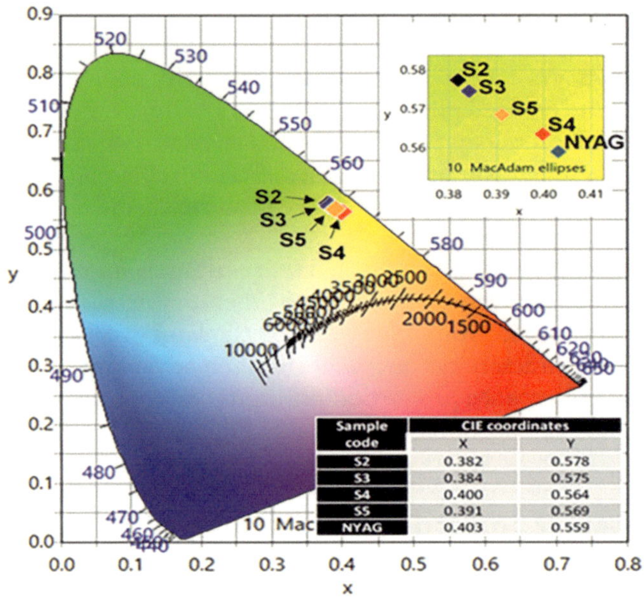

Figure 3.
CIE chromaticity diagram for different annealing treated YAG:Ce NPs and commercial NYAG. The inset shows the enlarged view on the CIE coordinates for each sample.

Additionally, the Commission internationale de I'Eclairage (CIE) chromaticity diagram was used to visualize the variation in emission properties for the YAG:Ce NPs under investigation. The CIE color coordinates of the samples, as shown in **Figure 3**, are in the greenish-yellow region, which is consistent with the emission color characteristic of YAG:Ce phosphor. It should be noted that S4 color coordinates were extremely similar to those of commercial YAG: Ce (NYAG4056-01-1, Intematix), indicating the usefulness of the synthesized phosphor. A notable observation was the linear shift in CIE color coordinates towards the deeper yellow region with an increase in CIE x and a decrease in CIE y coordinates. This shift is probably caused by the red component being more enriched from the broadened FWHM of the PL emission band. The ground state may not have fully split, resulting in a smaller energy difference between the $4f^1$ and 5d, as the PL emission FWHM narrowed with decreasing powder crystallinity [25, 26]. In addition, for noting: a.u. on the spectrum represent Arbitrary Scale (represented as arb. Units or a.u.).

5.1 Electroluminescence properties of YAG doped with cerium compound (Ce^{3+} ion)

To evaluate the suitability of the WLEDs made from various annealing-treated YAG:Ce NPs as white light emitters, photometric measurements were carried out on the devices. The obtained result was consistent with the PL analysis, indicating that enhanced powder crystallinity and stronger Ce^{3+} incorporation worked together to produce higher yellow fluorescence, which was the cause of the increasing tendency. Furthermore, S5 amply demonstrated the negative impact of cerium oxidation ($Ce^{3+} \rightarrow Ce^{4+}$), as the concentration of optically inert Ce^{4+} ions increased and consequently, the intensity of the yellow emission decreased. Consequently, the CIE chromaticity diagram (**Figure 4b**) was used to plot the corresponding CIE

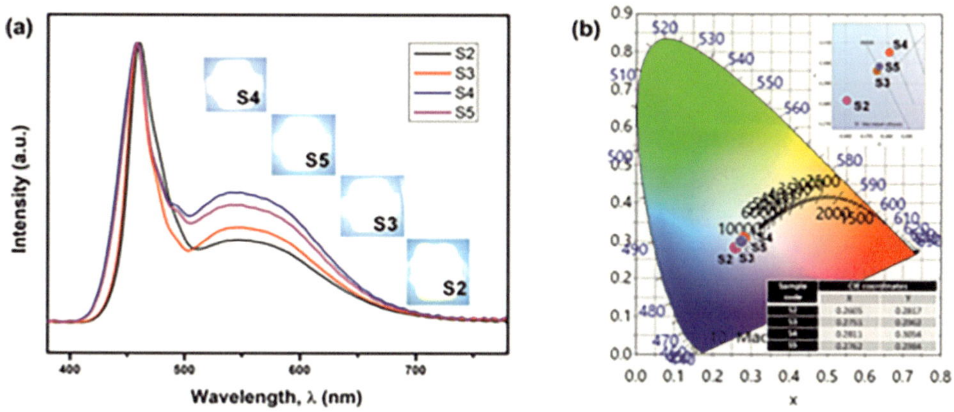

Figure 4.
EL spectra (a) and CIE coordinates (b) for the packaged WLEDs under a forward bias current of 20 mA. The inset of (a) and (b) each showed the photographic image of working WLEDs and a section of the 1931 CIE diagram, respectively.

color coordinates to represent the variations in the EL spectra of the WLEDs. For S2, the color coordinates were first placed in the bluish-white area. For S3, S5, and S4, the color coordinates moved linearly in the direction of the ideal white illumination (0.333, 0.333) along the blackbody locus. The results, which display a shift in color from bluish white to light cool-white, are in good agreement with the images captured on WLEDs. Following measurements, **Table 1** presents the results of the luminous efficacy of radiation (LER), purity, correlated color temperature (CCT), and color rendering index (CRI).

The outcome was a manifestation of higher yellow light output due to increased absorption of blue light, bringing the light output closer to the ideal white illumination. The S5 WLED's CRI, on the other hand, was comparable to the S4's, but because of the former's reduced yellow-to-blue light ratio, its CCT has significantly increased. Moreover, it was crucial to maintain the WLED's color purity at a minimum achievable value of 0.1 for perfect white illumination. **Table 1** shows that, among the fabricated samples, S4 WLED with the lowest purity indicated the highest optical quality based on its color coordinates that were closest to the ideal white illumination. The degree to which the phosphor emission spectrum satisfies the sensitivity of the human eye is indicated by the luminous efficacy of radiation (LER) [27].

Sample code	CCT (Kelvin, K)	CRI	Purity %	LER (lumen/watt, lm/w)
S2	13,370	79	29.12	261
S3	10,152	81	22.81	278
S4	9142	83	20.00	285
S5	9918	83	22.2	279

Table 1.
EL parameters of WLEDs fabricated with different annealing treated YAG:Ce NPs.

6. Photoluminescence properties of NMSO doped with cerium compound (Ce^{3+} ion)

The PLE spectra of the phosphors NMSO:$x$$Ce^{3+}$ that are monitored using the emission peak value are displayed in **Figure 5a**. The PL spectra of the phosphors excited by the excitation peak value are displayed in **Figure 5b**. The findings show that when near-UV light is applied to the phosphors, they display an asymmetric blue emission band. Because of the spin-orbit coupling of the 4f energy level and the 5d-4f transition, the asymmetric emission band is typical for Ce^{3+} doped phosphors.

The emission intensity and Ce^{3+} content are plotted in **Figure 5b**. From 0.001 to 0.005, the intensity rises with Ce^{3+} content before falling as the content rises even higher. Emission intensity is highest for NMSO:0.005 Ce^{3+}. The positions of the excitation and emission bands shift along with the increase in Ce^{3+} content, as does the emission intensity. As the Ce^{3+} content rises, there is a red shift seen in both the PLE and PL peaks. **Figure 5c** normalized excitation and emission spectra make it evident how the PLE and PL bands change as the Ce^{3+} content rises.

The PLE and PL peak values as a function of the Ce^{3+} content are also displayed in **Figure 5d**. As the Ce^{3+} content rises from 0.001 to 0.020, the excitation peak shifts from 339 to 356 nm and the emission band shifts from 404 to 441 nm. For x = 0.025, there is also a slight blue shift of the PLE and PL bands. Figure S3 displays the PLE and PL spectra of $Na_{2-x}MgSiO_4$:x Ce^{3+} for comparison. As the Ce^{3+} content increases, a red shift of the excitation and emission bands is also seen in $Na_{2-x}MgSiO_4$:x Ce^{3+}.

As shown in **Figure 5d**, the red shift's extent is, nevertheless, less than that of NMSO:x Ce^{3+} with Li^+. In addition to the asymmetric blue emission band, a shoulder

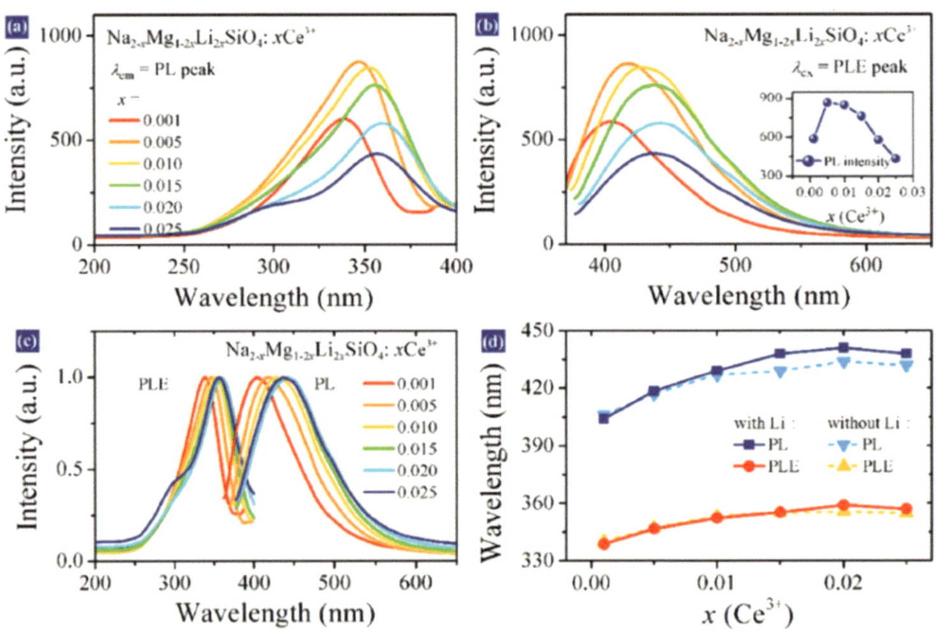

Figure 5.
(a) PLE spectra of $Na_{2-x}Mg_{1-2x}Li_{2x}SiO_4$:X Ce^{3+}; (b) PL spectrogram, inset: The variation trend of PL peak with Ce^{3+} content (x); (c) normalized PL and PLE spectra; (d) the variation trend of PL and PLE peak values with Ce^{3+} content.

is present at roughly 500 nm. The presence of two Na sites is suggested by the presence of a second environment for Ce^{3+}, as indicated by the shoulder at 500 nm.

6.1 Thermal stability and quantum efficiency of the phosphor

Figure 6a displays the PL spectra of NMSO:0.005 Ce^{3+} as a function of temperature (excited at 350 nm). It is evident that, as the temperature rises, the sample's emission intensity progressively drops while the emission peak essentially stays constant. The normalized integrated emission intensity as a function of temperature is displayed in **Figure 6b**.

The phosphor at 150°C maintains 57.6% of its intensity at room temperature because of thermal quenching. The data were examined using the Arrhenius equation to determine the relationship between temperature and luminescence intensity.

Figure 6b displays the fitted curve in accordance with. A value of 0.16 eV is the obtained E_a. Furthermore, NMSO:0.005 Ce^{3+} exhibits an absolute quantum efficiency of 86.0% when excited at 365 nm; corresponding spectra are displayed in Figure S5. The phosphor's efficiency and thermal stability imply that it is appropriate for use with PC-LEDs.

6.2 Electroluminescence properties of YAG doped with cerium compound (Ce^{3+} ion)

To assess the phosphors' potential for use, a 365 nm UV LED chip is assembled with commercial green-emitting $(BaSr)_2SiO_4:Eu^{2+}$, red-emitting $CaAlSiN_3:Eu^{2+}$, and NMSO:0.005 Ce^{3+} to create a white-emitting pc-LED. The electroluminescence (EL) spectra of a white PC-LED driven by various forward bias currents (50–350 mA) are displayed in **Figure 7a**. The entire visible light spectrum is covered by the EL band, and the normalized EL spectra at various currents almost exactly overlap. As illustrated in **Figure 7b**, the associated International Commission on Illumination (CIE) coordinates likewise almost overlap.

These phenomena imply that the phosphors emission positions are stable. As the working current increases, the temperature of the LED chip rises, a phenomenon that further suggests the thermal stability of the as-synthesized phosphor. The white PC-LEDs CRI value is 93.3 at 50 mA, and **Figure 7b** lists the PC-LED's other

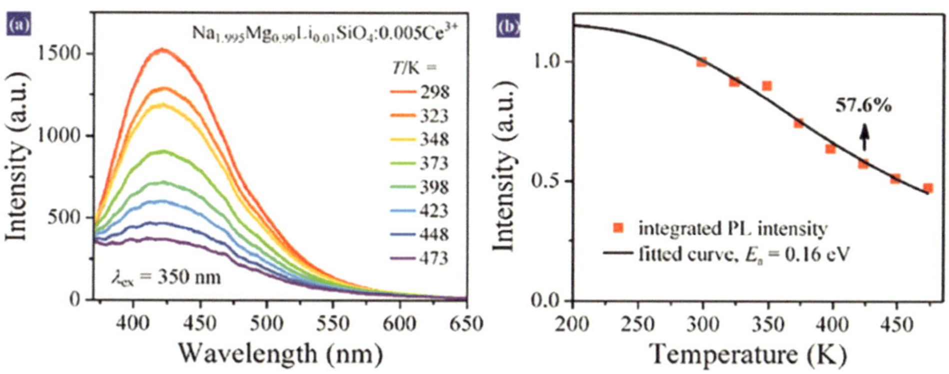

Figure 6.
(a) Temperature-dependent PL spectra of NMSO:0.005 Ce^{3+}, (b) normalized integrated luminescence intensity as a function of temperature and fitted curve.

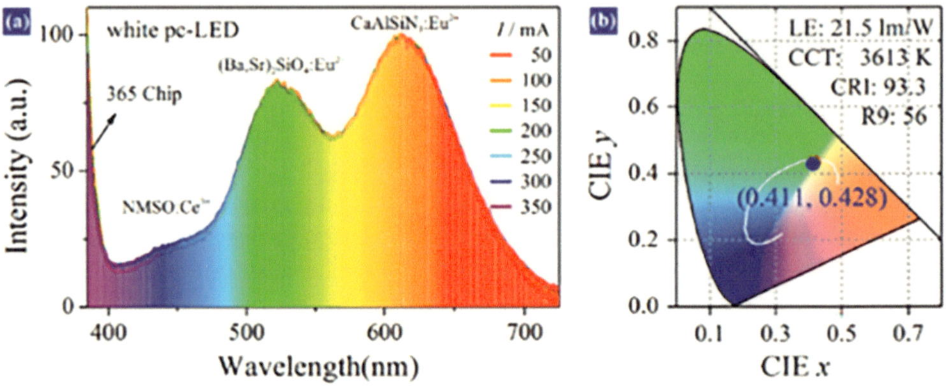

Figure 7.
(a) EL spectra of white pc-LED driven by different forward bias currents (50–350 mA), (b) parameters of the white pc-LED at 50 mA.

specifications. Based on these findings, it is possible to use the prepared phosphors as a potential blue-emitting phosphor for white PC-LEDs [28].

7. Applications of Ce^{3+} ions in LED technology

In the market today, several types of light-emitting diodes (LEDs) are designed for specific applications. One category is the traditional LEDs, which are primarily composed of inorganic materials. Another category is the Organic Light-Emitting Diodes (OLEDs), which are composed of organic materials. In traditional LEDs, light emission results from the recombination of electrons and holes within the band gap of inorganic materials, while in contrast to LEDs, OLEDs rely on the recombination process within organic materials, leading to the emission of light. In both cases, organic and inorganic materials are dependent on the hosts and dopants used.

7.1 Usage of Ce^{3+} compounds in LEDs and OLEDs for displays

Cerium-activated phosphor host materials play an integral part in LED lighting technology. This is achieved through the illumination of specific colors via concentration variations or through co-doping with other rare earth ions when excited by UV or blue light. Although the first invention of LED technology produced red light, Ce^{3+} doped aluminum garnet (YAG: Ce) played a pivotal role in producing white light of a more efficient nature. In particular, the combination of a blue-emitting phosphor and a yellow-emitting YAG: Ce^{3+} yielded white light of a superior nature [29]. The emission that originates from Ce^{3+} ion is due to the allowed electric dipole 5d → 4f transition, which produces a broad-band emission [29]. In this instance, when blue light or UV light excites a YAG: Ce^{3+} phosphor, produces yellow light. A combination of this yellow light with the remaining blue light results in a high-resolution of white light [30]. Basically, the color-correlated temperature (CCT) values change which subsequently produces a range of visible colors, such as different shades of white light (warm or cool), which may be applicable to various environmental settings [31].

On the other hand, the development of OLEDs for use in displays and illumination is a growing market. Lanthanides, specifically Ln^{3+}, are being explored in OLED technology for their sharp emission resulting from ^{4}f-^{4}f transitions [32]. Generally, obtaining blue emissions is a challenge; however, Tm^{3+} rare-earth ions are common due to their $^1G_4 \rightarrow {}^3H_6$ transitions, and our common Ce^{3+} ions are known for their allowed $^5d \rightarrow {}^4f$ transition [32]. While blue emissions from Ce^{3+} ions are rare in OLEDs, ongoing research in this field aims to tailor Ce^{3+} ions for that purpose. Alternatively, Ce^{3+} ions are used to excite other compounds [33], contributing to the efficiency of display devices. Research, such as that conducted by [34], demonstrates that Ce^{3+} ions yield optimal luminescence for a particular concentration. Thus, the use of Ce^{3+} ions in electroluminescent devices are mainly associated with enhancing organic emitters or exciting phosphors in organic materials through its UV emissions [32].

7.2 Incorporating Ce^{3+} compounds in energy-efficient lighting systems

Following the blue emissions discussed earlier, the most common method of achieving white light emissions is by combining a blue light-emitting chip, known as the indium gallium nitrate chip (InGaN), with a yellow-emitting phosphor material ($Y_3Al_5O_{12}:Ce^{3+}$), comprised of Ce^{3+} ions as a dopant, to produce the desired white light in lighting systems [35]. Despite these advances, its application remains limited due to low chemical stability and low luminous efficiency. Current research is focused on developing red-emitting phosphors with high luminous efficiency, excellent thermal stability, and high color purity [35] to produce a better quality of white light in display systems.

7.3 Future prospects and improvements in lighting and display systems

Although most single-doped phosphor-converted LEDs (pcLEDs) containing YAG:Ce^{3+} phosphors are commonly used to generate white light when excited by a blue LED source, this fabrication encounters stability issues due to the absence of the red color component. This limitation hinders the commercial feasibility of such LEDs as sources of white light [36]. This approach is perceived as a trade-off between color retention, Low Correlated Color Temperature (CCT) values, and luminous efficacy [36]. However, the main drawback is the absence of the red spectral component. The resulting white light typically has a CCT value exceeding 4000 K (cool white) and a low CRI (<80) [36]. In certain environments, such as indoor settings, a warm white light WW (2700-3200 K) is more preferred. Future advancements in this regard may involve a two-phosphor fabrication approach to produce white light—a combination of a yellow-emitting (YAG:Ce^{3+}) phosphor and a red-emitting phosphor material [35, 36].

7.4 Solar energy conversion

Solar energy conversion is a ground-breaking technology that harnesses solar power to generate electricity. It offers a sustainable and renewable source of energy, reducing greenhouse gas emissions and dependence on non-renewable resources. Solar energy conversion involves the process of transforming sunlight into usable energy. The primary method utilized for this purpose is by photovoltaic (PV) cells, also known as solar cells. These cells are typically made from semiconductor materials and can convert sunlight into electricity. When sunlight strikes the solar cells, it excites the electrons within the material, causing them to flow and generate an electric current. This direct current (DC) is then converted into alternating current (AC) by an inverter, making it suitable for powering appliances and feeding into the electricity grid [37].

7.5 Role of Ce^{3+} compounds in solar cells

Ce^{3+} compounds are also used in the application of solar cells. In this respect, crystalline silicon solar cells (PV module) can be applied to convert solar energy from the sun to electricity [Xue] in alignment of 4 different technologies, namely, down conversion, anti-reflective coatings, enhancement of electron transport properties, and enhanced stability.

i. **Down conversion**: To improve the efficiency of silicon solar cells, compounds such as Y_2SiO_5:Ce^{3+}, Yb^{3+} are mostly used because Yb^{3+} is well suited for energy excitation range (10,000 cm^{-1}) from ground state to the highest excited state, which corresponds to NIR emission of around 980 nm [37, 38]. In this compound, it can down-convert an excitation wavelength of 457 nm (blue violet) to 980 nm (NIR). Further, for this sample, Ce^{3+} can capture a broader range of UV light to cause emission of low energy photons, thereby making the solar cells very adaptable for this type of capture [38]. This process enhances the efficiency of silicon solar cells.

ii. **Anti-reflective coatings**: CaF_2:Ce^{3+}, Yb^{3+} are excellent luminescent materials that can be used in silicon solar cells as anti-reflective coating materials [37]. The purpose of this is to prevent the reflection of light to avoid damage to silicon materials and maximize efficiency, by using the CaF_2 material. For this process, high energy UV photons are converted to NIR photons through dopants (Ce^{3+}, Yb^{3+}) and the protective materials, enhancing efficiency. This allows more photons to penetrate the surface, hereby minimizing energy loss from the material [37].

iii. **Enhancement of electron transport properties:** In this process, Ce^{3+}-doped inorganic materials, such as SrF_2 in dye-sensitized solar cells, are used to convert high-energy photons to low-energy photons in the visible region through a process called down-shifting. This down-shifting process facilitates the transport of electrons, leading to an enhancement of energy and power efficiencies and thereby improving the overall performance of silicon solar cells [39].

iv. **Enhanced stability:** The enhanced stability of perovskite solar cells can be achieved by doping perovskite materials ($MAPbI_3$) with Ce^{3+} ions. The doping of Ce^{3+} ions lead to enhanced stability of the material, arising from the formation of a Ce^{3+}/Ce^{4+} redox pair and unique absorption in the UV region [40]. This leads to a more durable silicon solar cell that can withstand harsh environmental conditions.

7.6 Improving efficiency and stability of solar cells

Issues such as efficiency (as discussed) and stability are critical considerations for the commercialization of any of the above solar cells. Perovskite solar cells, in particular, have emerged as promising candidates in this regard, offering cost-effective fabrication, optimal performances at low temperatures, and high efficiency as both light absorbers and charge transporters [41]. These solar cells offer more than 20% efficiency, beating the 3rd generation of solar cells in their class. Coating attempts to address these issues often result in higher fabrication costs and make it heavier, thus making it an impractical solution. In an ideal situation, a solar cell should last for over 1000 hours, be stable at high temperatures, and have a lifespan of more than 20 years [41].

7.7 Challenges and future developments of the use of Ce^{3+} in solar cells

Since weather conditions play a major role in solar cell technology, integrating storage batteries could be advantageous for energy storage in the absence of sunlight [42]. The current global supply of electricity from solar PV generation is about 3%, and the intention is to increase this supply by 13% in 2030 and 25% in 2050 [42]. However, for this to be feasible, solar systems need to be economically viable when taking into consideration that commercially produced solar cells have efficiencies between 12 and 30% [42]. Alternatively, the sun's energy concentration can be directed to systems where collection of it is then transferred to heat fluids that will produce steam, which would then be able to turn turbines to produce electricity [42]. Ce^{3+} compounds are leading protagonists in such endeavors because they can collect UV rays and produce visible emissions through down-conversion processes. Such endeavors are achievable provided the efficiencies and stabilities of Ce^{3+} doped systems are improved.

8. Conclusion

These cerium compound possess significant potential in display technologies, lighting, and biological imaging applications. Recent research efforts focused on improving quantum efficiency and addressing the limitations associated with cerium compounds will further enhance their effectiveness. There are various synthetic methods that can be utilized to prepare phosphor materials doped with cerium compound such as solid-state reaction and combustion method. In this chapter, the comparison between air annealing and vacuum annealing for YAG:Ce NPs produced by combustion synthesis was observed on YAG:Ce NPs. The increase in PL emission intensity and a greater proportion of Ce3+ were the result of vacuum annealing. With longer annealing times, the NPs' CIE color coordinates are redshifted. The WLED made using YAG:Ce NPs and vacuum-annealed for five hours had the highest LER of 285 lm/W, according to the EL parameters, because the phosphor fluorescence and photopic curve overlapped more. The WLED emitted cool-white light could be further enhanced by fine-tuning the phosphor and packaging method. Na2MgSiO_4:Ce^{3+} study revealed that a high-temperature solid-state reaction was used to produce the phosphors. The phosphors emit blue light, which shifts the emission band to the longer wavelength side, when the host's Ce^{3+} content is increased or Li^+ ions are added. The reabsorption effect, energy transfer, and a stronger crystal field are the causes of this. Two Na sites harboring Ce^{3+} ions are responsible for the broad asymmetric emission band. The phosphor has a respectable level of thermal stability in addition to a high absolute quantum efficiency. The performance of the phosphor as-prepared suggests that PC-LEDs could use it as a possible material for white PC-LEDs.

Acknowledgements

Support from Mr. Magasi Emmanuel and Dr. Kurma Sunil for assisting in putting together the chapter.

Conflict of interest

The authors state that none of the work described in this chapter appears to have been influenced by any known competing financial interests or personal relationships.

Notes/thanks/other declarations

No additional declarations are available for this chapter.

Author details

Machaba L.A.Letswalo* and Leelakhrishna Reddy
Department of Physics, University of Johannesburg, Doornfontein, South Africa

*Address all correspondence to: letswalom@uj.ac.za

References

[1] Cerium Summary. Retrieved December 1, 2023, Available from: www.britannica.com/summary/cerium

[2] Cerium. Retrieved December 1, 2023, Available from: www.sciencedirect.com/topics/engineering/cerium

[3] Cerium Compounds. Retrieved December 1, 2023, Available from: www.en.wikipedia.org/wiki/Cerium_compounds

[4] Cerium. Retrieved December 1, 2023, Available from: www.en.wikipedia.org/wiki/Cerium

[5] Cerium. Encyclopædia Britannica, Encyclopædia Britannica, inc., Available from: www.britannica.com/science/cerium [Accessed: December 2, 2023]

[6] Optical and Scintillation Properties of Cerium-Doped $LaCl_3$, $LuBr_3$ and $LuCl_3$. Retrieved December 1, 2023, Available from: www.sciencedirect.com

[7] Journal of the Optical Society of America. Retrieved December 1, 2023, Available from: www.opg.optica.org/abstract.cfm?uri=josa-59-1-60

[8] Cerium Oxide Nanoparticles: Properties, Biosynthesis and Biomedical Application. Retrieved December 1, 2023, Available from: www.ncbi.nlm.nih.gov/pmc/articles/PMC9055511/

[9] Mokkelbost T, Kaus I, Grande T, Einarsrud MA. Combustion synthesis and characterization of Nanocrystalline *CeO2-based powders*. Chemistry of Materials. 2004;*16*(*25*):*5489-5494*

[10] Xianghua Y, Li F, Ye X, Xin X. Synthesis of cerium(iv) oxide ultrafine particles by solid-state reactions. Journal of the American Ceramic Society. 2000;**83**(4):964-966

[11] Souza LF, Vidal RM, Souza SO, Souza DN. Thermoluminescent dosimetric comparison for two different MgB_4O_7: Dy production routes. Radiation Physics and Chemistry. 2014;**104**:100-103

[12] Altunal V, Abusaid W, Guckan V, Ozdemir A, Yegingil Z. Luminescence characterization of Ce and Gd doped MgB_4O_7 phosphors. Journal of Luminescence. 2022;**246**:118815

[13] Ozdemir A, Altunal VO, Guckan V, Kurt K, Yegingil Z. Luminescence characteristics of newly-developed MgB_4O_7: Ce^{3+}, Na^+ phosphor as an OSL dosimeter. Journal of Alloys and Compounds. 2021;**865**:158498

[14] Santos CC, Valença JV, d'Errico F, Machado R, Caldas LV, Souza SO. Effect of different solvents on the optically stimulated luminescence signal from MgB_4O_7: Ce. Li-loaded polymer films. Radiation Protection Dosimetry. 2022;**198**(16):1230-1237

[15] Souza LF, Novais AL, Antonio PL, Caldas LV, Souza DN. Luminescent properties of MgB_4O_7: Ce, Li to be applied in radiation dosimetry. Radiation Physics and Chemistry. 2019;**164**:108353

[16] Souza LF, Silva AM, Antonio PL, Caldas LV, Souza SO, d'Errico F, et al. Dosimetric properties of MgB_4O_7: Dy, Li and MgB_4O_7: Ce, Li for optically stimulated luminescence applications. Radiation Measurements. 2017;**106**:196-199

[17] Gustafson TD, Milliken ED, Jacobsohn LG, Yukihara EG. Progress and challenges towards the development

of a new optically stimulated luminescence (OSL) material based on MgB_4O_7: Ce, Li. Journal of Luminescence. 2019;**212**:242-249

[18] Clayton RK. Light and Living Matter: A Guide to the Study of Photobiology. vol. 1: The physical part. 148 Seiten, 55 abb. und 2 tab. vol. 2: The biological part. 242 Seiten, 71 ABB. New York: McGraw-Hill Book Company; 1971. Preis: Vol. 1: 12,40 dm, vol. 2: 16,60 DM. Food / Nahrung. Jan 1972;**16**(7):810-811. DOI: 10.1002/food.19720160717

[19] Quack M. Molecular quantum dynamics from high resolution spectroscopy and laser chemistry. Journal of molecular structure. 1993;**292**:171-195

[20] Li G, Tian Y, Zhao Y, Lin J. Recent progress in luminescence tuning of Ce^{3+} and Eu^{2+}-activated phosphors for pc-WLEDs. Chemical Society Reviews. 2015;**44**(23):8688-8713

[21] Qin X, Liu X, Huang W, Bettinelli M, Liu X. Lanthanide-activated phosphors based on 4f-5d optical transitions: Theoretical and experimental aspects. Chemical reviews. 2017;**117**(5):4488-4527

[22] Shrestha N, Vandenbroucke D, Leblans P, Yukihara EG. Feasibility studies on the use of MgB_4O_7: Ce, Li-based films in 2D optically stimulated luminescence dosimetry. Physics Open. 2020;**5**:100037

[23] Liang J, Devakumar B, Sun L, Wang S, Sun Q, Huang X. Full-visible-spectrum lighting enabled by an excellent cyan-emitting garnet phosphor. Journal of Materials Chemistry C. 2020;**8**(14):4934-4943

[24] Guerbous L, Boukerika A. Nanomaterial host bands effect on the photoluminescence properties of Ce-doped YAG nanophosphor synthesized by sol-gel method. Journal of Nanomaterials. 2015;**16**(1):97-97

[25] Xia Z, Meijerink A. Ce^{3+}-doped garnet phosphors: Composition modification, luminescence properties and applications. Chemical Society Reviews. 2017;**46**(1):275-299

[26] Liang J, Sun L, Wang S, Sun Q, Devakumar B, Huang X. Filling the cyan gap toward full-visible-spectrum LED lighting with $Ca_2LaHf_2Al_3O_{12}$: Ce^{3+} broadband green phosphor. Journal of Alloys and Compounds. 2020;**25**(836):155469. DOI: 10.1016/j.jallcom.2020.155469

[27] Lau KS, Hassan Z, Lim WF, Quah HJ. Investigation on the effect of vacuum annealing time on structural and optical properties of YAG: Ce nanoparticles prepared by mixed-fuel microwave solution combustion synthesis. Optics & Laser Technology. 2022;**1**(154):108296. DOI: 10.1016/j.optlastec.2022.108296

[28] Chen M, Liao S, Han Y, Lian S, Zhang J. Color-tunable blue-emitting Na_2MgSiO_4: Ce^{3+} phosphors for white light-emitting diodes. Ceramics International. 2023;**49**(3):5431-5436. DOI: 10.1016/j.ceramint.2022.10.066

[29] Bachmann V, Ronda C, Meijerink A. Temperature quenching of yellow Ce^{3+} luminescence in YAG: Ce. Chemistry of Materials. 2009;**21**(10):2077-2084

[30] Islam NU, Usman M, Rasheed S, Jamil T. White light-emitting diodes: Past, present, and future. ECS Journal of Solid State Science and Technology. 2021;**10**(10):106004

[31] Sefage AP, Mamo MA, Masiteng PL, Balakrishna A, Coetsee E, Swart HC, et al. The concentration effect of Ce^{3+} ions on the photoluminescence properties of the host sensitized $NaMPO_4$-Ce^{3+} (M= Mg and Ca) phosphors via solid-state reaction method. Physica B: Condensed Matter. 2022;**15**(647):414383

[32] Monteiro JHSK, de Bettencourt-Dias A. Lanthanide ion emission in multicolor OLEDs (Ce^{3+}, Pr^{3+}, Tb^{3+}, Dy^{3+}, Tm^{3+}, and white light Eu^{3+}/Tb^{3+} hybrid systems) and device characterization. In: Lanthanide-Based Multifunctional Materials. Elsevier; 2018. pp. 99-131. DOI: 10.1016/b978-0-12-813840-3.00003-x

[33] Yu T, Su W, Li W, Hua R, Chu B, Li B. Ultraviolet electroluminescence from organic light-emitting diode with cerium (III)–crown ether complex. Solid-state electronics. 2007;**51**(6):894-899

[34] Zheng XL, Liu Y, Pan M, Lü XQ, Zhang JY, Zhao CY, et al. Bright blue-emitting Ce^{3+} complexes with encapsulating polybenzimidazole tripodal ligands as potential electroluminescent devices. Angewandte Chemie International Edition. 2007;**46**(39):7399-7403

[35] Qiao X, Zhao J, Xu C, Liu Y, Tsuboi T. Emission enhancement of Eu^{3+} ions in LiY $(PO_3)_4$ phosphate phosphors via anionic SO_4^{2-} substitution. ECS Journal of Solid State Science and Technology. 1 Jun 2023;**12**(6):066010. DOI: 10.1149/2162-8777/acdea4

[36] Meyer J, Tappe F. Photoluminescent materials for solid-state lighting: State of the art and future challenges. Advanced Optical Materials. 2015;**3**(4):424-430

[37] Liu X, Hu X, Miao H, Zhang G, Mu J, Han T, et al. CaF_2: Ce^{3+}/Yb^{3+} hollow spheres luminescence downconversion property optimize anti-reflective coatings for solar cells. Solar Energy. 2016;**134**:45-51

[38] Xue H, Xu Z, Zhang M, Wang J, Ruan W. Ethylene vinyl acetate films filled with ytterbium containing rare earth particles (Y_2SiO_5: Ce^{3+}, Yb^{3+}) which have optical down-conversion capabilities and useful for encapsulating solar cells. Journal of Plastic Film & Sheeting. 2015;**31**(3):233-247

[39] Samuel AK, Dileep KR, Jayaraman M, Veerappan G, Jayaraman S. Enhanced power conversion efficiency using a Ce^{3+}: SrF_2 Down-shifting nanophosphor-based Photoelectrode for dye-sensitized solar cell applications. ACS Applied Energy Materials. 2021;**4**(7):7112-7121

[40] Song Z, Xu W, Wu Y, Liu S, Bi W, Chen X, et al. Incorporation of lanthanide ions into perovskite film for efficient and stable perovskite solar cells. Nano-Micro. 2020;**16**(4):2001770

[41] Tiep NH, Ku Z, Fan HJ. Recent advances in improving the stability of perovskite solar cells. Advanced Energy Materials. 2016;**6**(3):1501420

[42] Shahabuddin M, Alim MA, Alam T, Mofijur M, Ahmed SF, Perkins G. A critical review on the development and challenges of concentrated solar power technologies. Sustainable Energy Technologies and Assessments. 2021;**1**(47):101434

Chapter 3

Effect of Silver Nanoclusters on Physical and Electrical Properties of Cerium Oxide Thin Film

Mousri Paul, Sabyasachi Karmakar and Supratic Chakraborty

Abstract

An *in situ* X-ray photoelectron spectroscopy (XPS) investigation has revealed that silver metal clusters (Ag-NCs) can enhance the redox property of cerium oxide (CeO_2) at relatively lower temperatures by oxidizing Ag-NCs to Ag^{2+} and Ag^{3+} states. Strong metal support interaction (SMSI) effect at the interface is indicated by a specific interaction seen in high-resolution transmission electron microscopy (HRTEM) images of CeO_2/Ag-NCs, confirming Ag-NC encapsulation by deposited CeO_2 layer after heating. Through chemisorption processes, the SMSI effect aids in the release of oxygen from the ceria surface by making a bond of Ag, O, and Ce. Consequently, employing silver as a supporting novel metal improves the redox behaviour of CeO_2 at nearly 100°C. The band gap of cerium is modified because of this interaction as shown by UV-vis spectroscopy, influencing the electronic charge transport property of ceria. The current-voltage (I-V) characteristics in silver cluster supported ceria thin film verify the significant increase in current under visible light illumination as compared to the current in dark conditions. This renders that Ag-NCs supported CeO_2 is suitable for photocatalysis and the capacitance-voltage (C-V) characteristic confirms the enhanced storage capacity of Ag-NCs /CeO_2-based metal-oxide-semiconductor (MOS) devices qualifies it for use as non-volatile memory (NVM) devices.

Keywords: thin film of cerium oxide, silver cluster supported ceria, low temperature redox reaction, response under visible light, NVM memory device

1. Introduction

Cerium metal is one of the most interesting element in Lanthanum group of periodic table as it shows different physical properties in oxidized form. It is well known that cerium metal ([Xe]$4f^15d^16s^2$; four outer valance electrons) is reactive in air and changes to CeO_2 with electronic configuration [Xe]. Depending on the presence of valance electrons the electronic structure of cerium vary as a result the cerium present in mixed valance state. This property makes this material an important place in industry for various applications as heterogeneous catalyst for oxidizing hazard gasses [1], water gas splitting [2], in memristors for application in meromorphic computing [3] also as a photodiode [4]. Minimization of redox reaction temperature of cerium oxide to make it useful as heterogeneous catalyst is one of the most challenging work now a days. Several studies have done where it was clearly indicated that the chemical state of ceria is most important factor to improve the catalytic behavior of ceria [5]. In general the bulk cerium oxide started redox reaction after heating at 600°C [6].

The most important nature of cerium oxide is that its oxygen adsorption and desorption property based on the experimental environment. Researchers were trying to go in a conclusion of that procedure by many ways but not succeed till now. Most of researchers are believe that cerium oxide changes its oxidation state by following Mars-van Krevelen (MvK) mechanism when one oxygen ion remove from the surface of ceria then two of neighboring Ce^{4+} ions take the electrons and form Ce^{3+} ions to make charge neutrality [7]. To occur this reaction we need to control some parameters during the formation of ceria.

Introducing metals, the redox reaction temperature improved and occur at relatively lower temperature has been studied by many researchers [8–12]. Mostly the novel metals give the better result in that case compare to other metals. The growth condition of the sample determines the interaction between ceria and the corresponding metals. Many research on different structure of supported metals as single atom [13], cluster of metals [14], doping with ceria [15] has been studied for getting better redox reaction at relatively lower temperature. In general cerium oxide having higher oxygen storage and release capacity when it present in mixed valance state (presence of Ce^{3+} and Ce^{4+} state together). Introducing metals, an increment in the oxygen storage capacity within ceria occur due to the interaction at both the surface and interface between metals and ceria. This interaction is known as the strong metal support interaction (SMSI) which was first observed by Tauster et al. on TiO_2 [16]. In 1990, research on increasing oxygen storage capability of ceria by introducing rare earth metal was described by Miki et al. and Cho by considering the two types of lattice defects within ceria [17]. The SMSI effect at the interface takes an important role in redox property of ceria.

The interaction and the corresponding structural change of cerium oxide with the supported metals are different in case of bulk as compare to the thin film of ceria. As it is already known that the formation of a thin film of a material from bulk can change the dimension of the material i.e. the material completely changes from 3D to 2D structure in nature. As a result the physical properties of the 2D structure is different from 3D one which affects direct on the electron transport of the material. Therefore, people are growing their interest on controlling the redox reaction of cerium oxide based thin films to make it applicable in form of devices. Considering the supported metal, as single atom and cluster or nanoparticles of a metal shows different properties. In some metals the cluster formation of metal changes its electronic structure.

Considering this situation, we have investigated the interaction between novel metal (silver Ag) cluster with cerium oxide thin film by varying temperature from 25 to 250°C. Using this combination we have achieved the redox reaction temperature at 100°C as verified from the results of in-situ XPS experiment. Based on the results we have prepared one metal-oxide-semiconductor (MOS) structured based device to get to know the effect of Ag-metal on the electrical transport of ceria to application in non-volatile memory devices. Presence of silver metal clusters within cerium oxide also affects the optical band gap and from current-voltage study of Ag-NCs supported ceria an optical response of the device we have achieved under visible light for application in photocatalysis.

2. Experimental details

We have prepared the cerium oxide thin film by rf magnetron sputtering technique using 99.999% pure cerium oxide target. The *p*-type silicon (100) was used as a substrate for this purpose. We have used common RCA cleaning method [18] to clean the substrate. During sputtering we have used rf power of 45 W with 5mTorr of deposition pressure. The deposition rate is maintained as 1 nm/min. Deposited cerium oxide having thickness of 10 nm on which the silver (Ag) metals in form of nanocluster of size 4 nm was deposited using NC200U cluster deposition UHV system which is equipped with a quadrupole mass filter and x-ray photoelectron spectroscopy (XPS) technique. To compare the silver cluster supported ceria thin film combination we have also deposited pristine ceria thin film using same deposition parameters. The physical characteristics of pristine and Ag-nanocluster/CeO_2 have performed by differential scanning calorimetry (DSC), cross sectional TEM study, in-situ XPS, UV-vis spectroscopy and grazing incidence X-ray diffraction (GIXRD) technique.

To form a MOS structure of this thin film combination we have again deposited cerium oxide capping layer of thickness 40 nm on Si/ CeO_2/Ag-NCs thin film. For electrode deposition we have patterned on the film using photomask of 100 μ diameter with a separation of 25 μ using photolithography technique. We have then used e-beam evaporator to deposit titanium metal of thickness 300 nm. The capacitance-voltage and current-voltage characteristics was performed using Agilent E4980A LCR meter and Keithley 2450 source-meter, respectively in signatron microprobe station with a heating facility of 25–300°C.

3. Experimental results and discussions

The effect of heating on pristine ceria and silver cluster supported ceria was verified by performing the DSC study is presented in **Figure 1(a)**. In case of pristine ceria thin film the endothermic peak appeared at 150 and 200°C because of the involving of *2nd* order phase transition of ceria. In silver cluster supported ceria thin film the exothermic peak appeared at 200°C indicating a phase transition with a corresponding endothermic peak at 595°C because of melting of Ag-cluster. As it is known that bulk silver having melting temperature at 961°C i.e. the formation of silver clusters makes a reduced melting temperature. **Figure 1(b)** and **(c)** are the SEM images of pristine ceria and after depositing Ag-NCs on ceria thin film.

GIXRD and TEM study has been performed at room temperature and after annealing under vacuum at 250°C [19]. From GIXRD study of pristine ceria and silver supported ceria the presence of silver is confirmed within the sample. A clear difference in the size of deposited silver clusters was observed after heating at 250° C as well as an encapsulation of clusters by ceria surface layer was also observed from cross-sectional TEM study. The HRTEM images clearly verifies a layer of ceria fully cover the deposited silver clusters during heat treatment whereas the as deposited silver clusters are attached on the deposited cerium oxide thin film at room temperature i.e. without performing any annealing.

Based on the DSC study the in-situ XPS of both the samples has been performed and presented in **Figures 2–5**. From the XPS curve of Cerium 3d in absence of presence of silver metal clusters it is clearly observe that the improvement of the

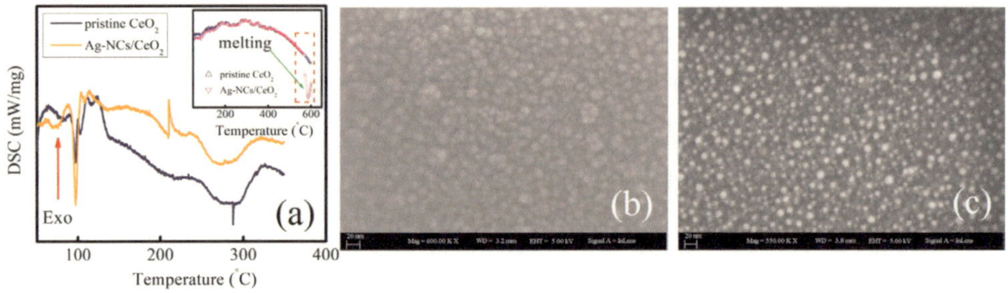

Figure 1.
(a) Is the DSC curve of pristine ceria and silver cluster supported ceria thin film, (b) is the SEM image of as deposited ceria thin film and (c) is the SEM image after depositing Ag-clusters on ceria.

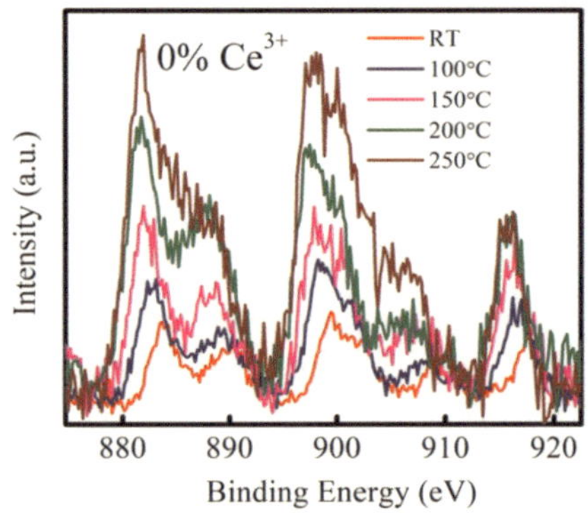

Figure 2.
Ce3d XPS spectra with temperature for pristine ceria thin film.

redox property of cerium oxide in silver cluster supported cerium oxide thin film. There is no effect of temperature on the redox property of ceria in case of pristine ceria whereas an improved redox property at relatively lower temperature as 100°C is observed from in-situ XPS study for silver supported ceria thin film. A detailed study on pristine ceria and silver supported ceria thin film is presented in [19] with detailed analysis and description. **Figure 4** is the XPS spectra of Ag3d where the reversible change of oxidation states of silver clusters was verified with increasing temperature. The interaction of cerium with silver via oxygen due to the chemisorption process is the reason of oxidation of silver which can help to make the reduction of ceria at relatively lower temperature. A lower binding energy shift of oxygen XPS spectra in Ag-Nc/CeO_2 sample as observed in **Figure 5** also indicates an involvement of Ce^{3+} states within the sample.

The summary of the experiment is that the presence of silver affects the redox property of ceria in form of thin film after heat treatment. And the required temperature is relatively low in that case which can make this combination useful for application in low temperature catalytic activity of ceria.

3.1 Effect of light on novel metals

Over the past few decades the use of metal nanoparticles (NPs) to utilize light absorption grabbing the attention. As absorption of light is an interesting property of

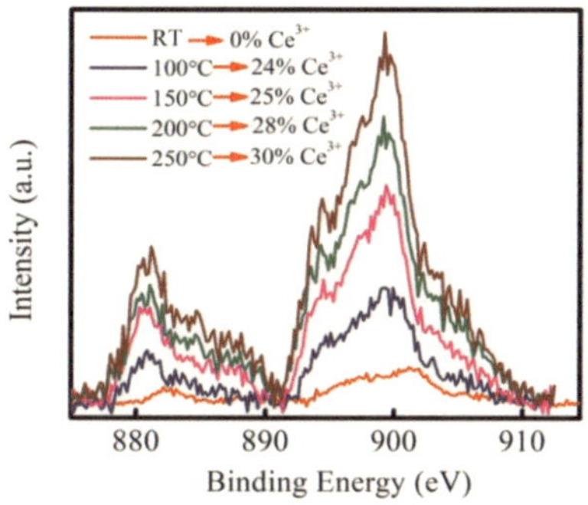

Figure 3.
Ce3d XPS spectra with temperature for Ag-NCs/ceria thin film.

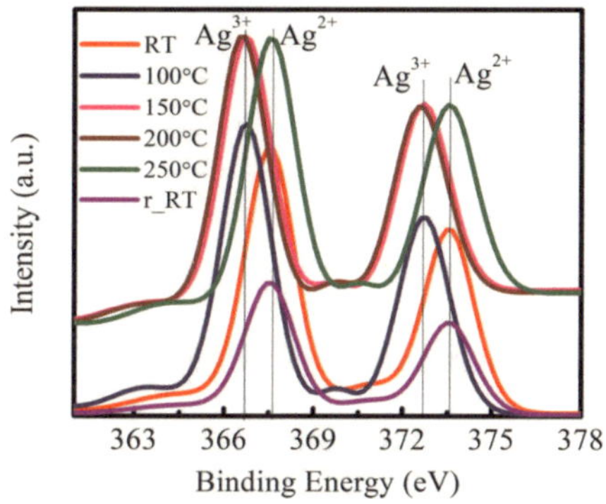

Figure 4.
Ag3d XPS spectra with temperature for Ag-NCs/ceria thin film.

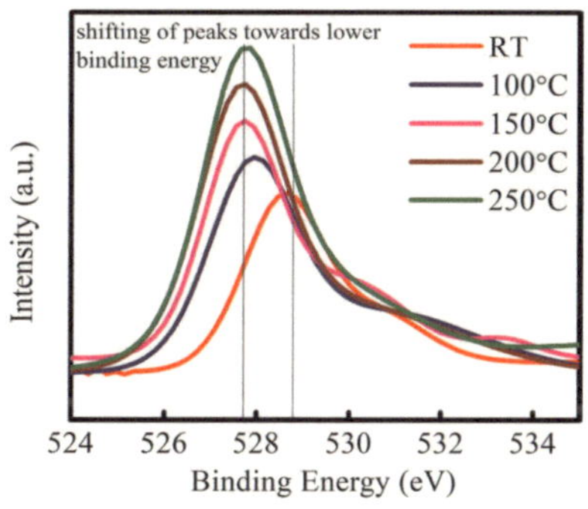

Figure 5.
O1S XPS spectra with temperature for Ag-NCs/ceria thin film.

metal NPs and by introducing this NPs within oxides it can be used as a photocatalysis which is known as metal-induced photocatalysis (MIP) [20]. The effect of light absorptions of metal NPs under illumination of visible light is one of the challenging work in photocatalyst study. The localized surface plasmon resonance (LSPR) and the interband transition mainly the reason for the generation of MIP process in metal NPs [21]. Mostly for novel metal NPs the LSPR absorption occur near the visible light region. The metal oxides mostly TiO_2 and CeO_2 having band gap energy fallen into

UV-region those are mostly inactive under visible light where the electron-hole pair generate only after UV illumination. By introducing novel metals within these two oxides the visible light activity of these oxides enhance. Jiang et al. observed that by introducing Ag in TiO_2 the photocatalytic activity become higher than TiO_2 under visible light [22] where the activity of novel metals with oxides highly depending on the size of metal NPs. Based on this we have performed the electrical transport study of silver-nanocluster supported cerium oxide thin film based MOS devices under visible light and dark condition.

3.2 Application in optoelectronics

A schematic diagram of MOS structure of Ag-NCs/CeO_2 combination is depicted in **Figure 6(a)**. The I-V curve of silver supported ceria thin film under dark and light condition is presented in **Figure 6(b)** under voltage sweep from 0 to 5 V and 5 to 0 V. The current density become increases under visible light environment as compare to the current under dark condition. Under the visible light illumination condition the Ag-NCs produces an electron-hole pair and after applying voltage the charge separation occur as a result the charge carrier density increasing more than that of dark condition. Therefore, the optical activity of silver supported ceria thin film under visible light appear i.e. this combination can be used as a photo catalysis.

3.3 Application as non-volatile memory devices

The novel metal nanoparticle embedded high -κ dielectric based non-volatile memory (NVM) devices gives better performance as faster write and erase capacity, longer retention time and better endurance by smaller applied voltage [23]. Metal nanoparticle with transition metal oxide composite in the application of NVM devices was studied very few within 2009–2016 year range where researchers use mostly novel metals (Ag and Pt) with mainly two oxides such as HfO_2 and Al_2O_3 [24, 25]. In these studies the effect of size of nanoparticles as well as the deposition conditions are the important factors for getting better storage devices. Inspired from this, we have used Ag-nano cluster supported cerium oxide combination based NVM devices. As it is good for us that ceria itself is known as higher oxygen reservoir by changing its oxidation states depending on the environmental circumstances. From other physical characteristics we have observed there is a change of oxidation state of ceria after

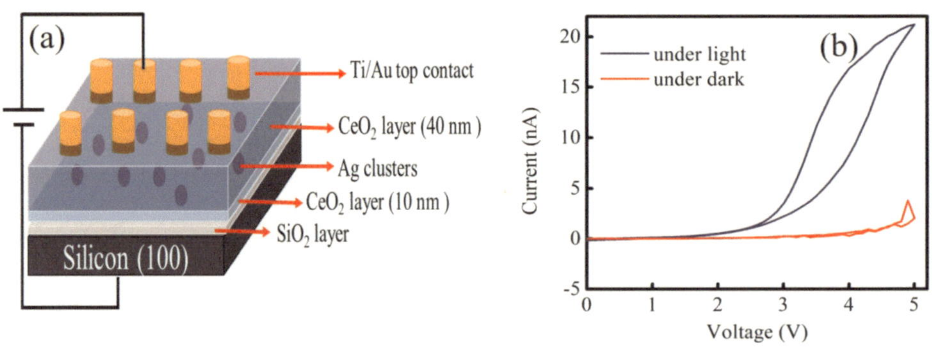

Figure 6.
(a) Is the schematic diagram of MOS based structure, (b) is the I-V curve of Ag cluster supported cerium oxide thin film under dark and illumination of visible light.

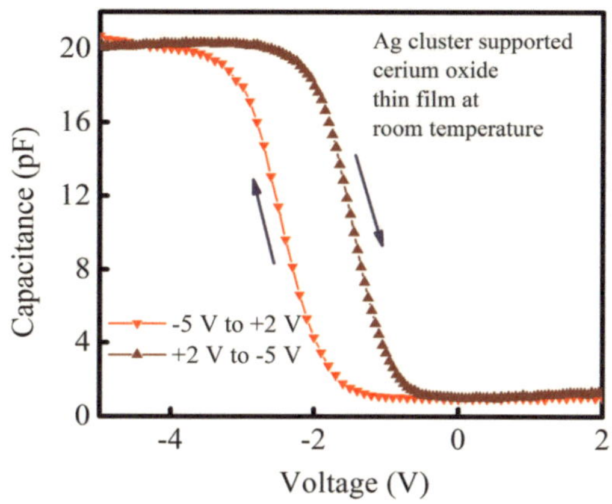

Figure 7.
High frequency C-V curve of as deposited Ag cluster supported cerium oxide thin film.

annealing at a certain temperature in presence of silver and with heating up better redox reaction we have achieved.

Figures 7 and **8** are representing the C-V characteristics of Ag-NCs/CeO_2 composite of as deposited film and after annealing at 250°C. It is clearly observed that as deposited Ag-cluster supported ceria thin film forming a hysteresis loop when the voltage sweeps from −5 to +2 V and + 2 to −5 V. After annealing at 250°C, hysteresis loop area become negligible because of the charge store in silver clusters by electron or hole injections [26]. From this experimental result it is concluded that presence of silver within ceria thin film makes the device performance better than pristine ceria and after annealing the device performance decreases.

4. Conclusions

We have studied the effect of interaction of silver metal in form of clusters of selected size with cerium oxide during heating and the corresponding change of electrical properties. Interaction of Ag-NCs with cerium oxide in form of thin film

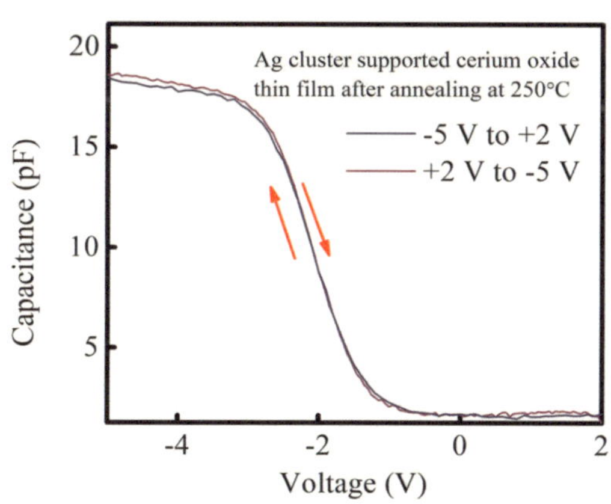

Figure 8.
High frequency C-V curve of Ag cluster supported cerium oxide thin film after heating at 250° C.

results the improvement of reduction temperature (at nearly 100°C) and can change the oxidation state of ceria from 4+ to 3+. The composition of Ag-NCs/CeO_2 thin films involve in the SMSI effect at the interface of metal and supported cerium oxide as a result metal clusters encapsulated by cerium oxide layer after heat treatment as observed from HRTEM images. The oxidized form of deposited silver clusters with an reversible change of oxidation states can helps to improve the redox property of ceria by involving interaction via chemisorption process is verified from XPS spectra of Ag and Ce3d. This change in physical properties also helps to improve the electron charge transport of silver supported ceria and makes this combination applicable as photo catalysis in form of device under visible light illumination.

Acknowledgements

Authors wish to acknowledge Mr. Debraj Dey for helping in XPS and SEM measurement and Ms. Dimitra Das, Department of Physics, Jadavpur University, Kolkata for UV-Vis diffuse reflectance measurement. The first author wishes to acknowledge the University Grant Commission (UGC), Govt of India for financial help.

Conflict of interest

The authors declare no conflict of interest.

Abbreviations

DSC	differential scanning calorimetry
GIXRD	grazing incidence X-ray diffraction
XPS	X-ray photoelectron spectroscopy
NVM	non-volatile memory
SMSI	strong metal-support interaction
MOS	metal-oxide-semiconductor
UHV	ultra high vaccum
TEM	transmission electron microscopy
HRTEM	high resolution transmission electron microscopy
LSPR	localized surface plasmon resonance

Author details

Mousri Paul[1,2*†], Sabyasachi Karmakar[1,2†] and Supratic Chakraborty[1]

1 Saha Institute of Nuclear Physics, Kolkata, India

2 Homi Bhabha National Institute, Mumbai, Maharashtra, India

*Address all correspondence to: mousri1993paul@gmail.com

† These authors contributed equally.

References

[1] Wu Y, Lin J, Ma G, Xu Y, Zhang J, Samart C, et al. Ni nanocatalysts supported on mesoporous Al_2 O_3-CeO_2 for CO_2 methanation at low temperature. RSC Advances. 2020; **10**(4):2067-2072

[2] Ramshanker N, Ganapathi KL, Bhat MS, Mohan S. Rf sputtered CeO_2 thin films-based oxygen sensors. IEEE Sensors Journal. 2019;**19**(22): 10821-10828

[3] Rehman S, Kim H, Khan MF, Hur J-H, Lee AD, Kee Kim D. Tuning of ionic mobility to improve the resistive switching behavior of Zn-doped CeO_2. Scientific Reports. 2019;**9**:19387

[4] Suresh R, Ponnuswamy V, Sankar C, Manickam M, Mariappan R. Influence of Co concentration on the structural, optical, morphological and photo-diode properties of cerium oxide thin films. Ceramics International. 2016;**42**: 12715-12725

[5] Matolin V, Sedlacek L, Matolinova I, Sutara F, Skala T, Smid B, et al. Photoemission spectroscopy study of Cu/CeO_2 systems: Cu/CeO_2 nanosized catalyst and CeO_2(111)/Cu(111) inverse model catalyst. The Journal of Physical Chemistry C. 2008;**112**:3751-3758

[6] Henderson MA, Perkins C, Engelhard M, Thevuthasan S, Peden C. Redox properties of water on the oxidized and reduced surfaces of CeO_2(111). Surface Science. 2003;**526**: 1-18

[7] Kullgren J. Oxygen Vacancy Chemistry in Ceria. Uppsala: Uppsala University, Structural Chemistry; 2012. p. 896

[8] Schierbaum K. Ordered ultra-thin cerium oxide overlayers on Pt(111) single crystal surfaces studied by LEED and XPS. Surface Science. 1998;**399**:29

[9] Berner U, Schierbaum K. Cerium oxides and cerium-platinum surface alloys on Pt(111) single-crystal surfaces studied by scanning tunneling microscopy. Physical Review B. 2002;**65**: 235404

[10] Eck S, Castellarin-Cudia C, Surnev S, Ramsey MG, Netzer FP. Growth and thermal properties of ultrathin cerium oxide layers on Rh(111). Surface Science. 2002;**520**:173

[11] Xiao W, Guo Q, Wang EG. Transformation of CeO_2(111) to Ce_2O_3(0001) films. Chemical Physics Letters. 2003;**368**:527

[12] Zhao X, Ma S, Hrbek J, Rodriguez JA. Reaction of water with Ce-Au(111) and CeOx/Au(111) surfaces: Photoemission and STM studies. Surface Science. 2007; **601**:2445

[13] Liu J-C, Wang Y-G, Li J. Toward rational design of oxide-supported single-atom catalysts: Atomic dispersion of gold on Ceria. Journal of the American Chemical Society. 2017;**139**:6190-6199

[14] Lin F, Hoang DT, Tsung C-K, Huang W, Lo SH-Y, Wood JB, et al. Catalytic properties of Pt cluster-decorated CeO_2 nanostructures. 2011;**4**: 61-71

[15] Khalakhan I, Dubau M, Haviar S, Lavkova J, Matolinova I, Potin V, et al. Growth of nano-porous Pt-doped cerium oxide thin films on glassy carbon substrate. Ceramics International. 2013; **39**:3765-3769

[16] Tauster SJ, Fung SC, Garten RL. Strong metal-support interactions.

Group 8 noble metals supported on titanium dioxide. Journal of the American Chemical Society. 1978;**100**: 170

[17] Miki T, Ogawa T, Haneda M, Kakuta N, Ueno A, Tateishi S, et al. Enhanced oxygen storage capacity of cerium oxides in $CeO_2/La_2O_3/Al_2O_3$ containing precious metals. ChemInform. 1990;**21**:6464-6467

[18] Kern W. Cleaning solutions based on hydrogen peroxide for use in silicon semiconductor technology. RCA Review. 1970;**31**:187-206

[19] Paul M, Satpati B, Chakraborty S. Interaction of silver nano-clusters with ceria thin-films: An in situ temperature dependent x-ray photoelectron spectroscopy study. Journal of Alloys and Compounds. 2022;**911**:164956

[20] Bumajdad A, Madkour M. Understanding the superior photocatalytic activity of noble metals modified titania under UV and visible light irradiation. Physical Chemistry Chemical Physics. 2014;**16**:7146-7158

[21] Liu L, Zhang X, Yang L, Ren L, Wang D, Ye J. Metal nanoparticles induced photocatalysis. National Science Review. 2017;**4**:761-780

[22] Murugadoss G, Kumar DD, Kumar MR, Venkatesh N, Sakthivel P. Silver decorated CeO_2 nanoparticles for rapid photocatalytic degradation of textile rose Bengal dye. Scientific Reports. 2021;**11**: 1080

[23] Naqi M, Kwon N, Jung SH, Pujar P, Cho HW, Cho YI, et al. High-performance non-volatile InGaZnO based flash memory device embedded with a monolayer Au nanoparticles. Nanomateals. 2021;**11**:1101

[24] Yun M, Mueller DW, Hossain M, Misra V, Gangopadhyay S. Sub-2 nm size-tunable high-density Pt nanoparticle embedded nonvolatile memory. IEEE Electron Device Letters. 2009;**30**:1362-1364

[25] Mukherjee S, Zheng H, Gangopadhyay K, Gangopadhyay S. Influence of Pt nanoparticle induced defects and surface coverage in determining asymmetric programming/erasing signatures for nanocrystal embedded nonvolatile memory applications. Advanced Materials Interfaces. 2016;**3**:1600436

[26] Biswas D, Mondal S, Rakshit A, Bose A, Bhattacharyya S, Chakraborty S. Size and density controlled Ag nanocluster embedded MOS structure for memory applications. Materials Science in Semiconductor Processing. 2017;**63**: 1369-8001

Chapter 4

Impact of Rare Earth Element Cerium on Structural and Magnetic Characterization of Different Nanoferrites

Kannan Elaya Kumar, S. Muthulingam and A. Manikandan

Abstract

Due to its numerous uses in the fields of magnetic recording devices, electronic devices, biosensors, high-frequency transformers, microwave devices, and radar devices, magnetic nanoferrites have drawn increased attention from society. There are several ways to make rare earth ion-doped spinel ferrites, including the sol-gel approach, auto-combustion method, hydrothermal method, co-precipitation method, ball milling method, and green synthesis method. Numerous studies have recently discussed the effect of rare earth ion substitution on spinel ferrite, concluding that it improved the material's structural and magnetic characteristics. The replacement of Cerium ions, which was studied by just a few researchers, allows ferrites and mixed ferrites' magnetic properties, such as coercivity, saturation magnetization, and retentivity, to be fine-tuned. In this study, we summarized the effect of different concentrations of Cerium ions on various ferrites to find the significant results of structural parameters like crystal size, lattice constant, and also magnetic parameters such as coercivity and saturation magnetization, through the study of X-ray diffraction (XRD) and vibrating sample magnetometer (VSM).

Keywords: Cerium, spinel ferrites, XRD, VSM, crystal size

1. Introduction

The rare earth elements (REEs) are otherwise called as rare earth metals, or they are referred to as lanthanides containing 17 indistinguishable heavy metals (La, Ce, Pr, Nd, Pm, Sm, Eu, Gd, Tb, Tm, Yb, Lu, Dy, Ho, Er). Apart from these rare earth elements (REEs), Scandium and Yttrium (Sc and Y) are also considered as rare earth elements of the lanthanide group, since it possesses chemical properties of Scandium and Yttrium that have different magnetic properties than the lanthanides. Normally, lanthanide groups are classified into light rare earth elements and heavy rare earth elements. The elements from Lanthanum to Europium are grouped under light rare earth elements, while the elements from Gadolinium to Lutetium are grouped under heavy rare earth elements. The physical and chemical properties of all lanthanide

Figure 1.
Symbol of Cerium.

elements are similar due to the progressively filled 4f orbitals with increasing atomic number. In that, ionic radius gradually decreases from La^{3+} (1.06 Å) to Lu^{3+} (0.85 Å), which is the so called the lanthanide contraction. The symbol of Cerium is shown in **Figure 1**.

The incorporation of rare earth elements, such as Cerium, into the structural matrix of nanoferrites has garnered considerable attention due to the profound influence these elements exert on both structural and magnetic characteristics. Rare earth elements possess unique electronic configurations and magnetic properties that can significantly alter the crystal lattice and magnetic behavior of nanoferrites. Among these, Cerium, a member of the lanthanide series, stands out for its versatile chemical nature and varied oxidation states. In this context, this introduction aims to explore the impact of Cerium doping on the structural aspects, such as crystal symmetry and lattice parameters, as well as the magnetic properties of different nanoferrites. Understanding the interplay between Cerium and nanoferrite matrices is crucial for tailoring and optimizing the performance of these materials in various technological applications, ranging from catalysis to magnetic devices. This exploration provides a foundation for delving into the subsequent sections, which delve into specific analyses and findings related to the structural and magnetic transformations induced by Cerium incorporation into different nanoferrites.

In the past few decades, REE has been applied in the field of various devices, such as rechargeable batteries, light-emitting diodes (LEDs), solar panels, magnetic resonance imaging (MRI), superconductors, supermagnets, catalysts, storage devices, glass additives, fluorescent materials, and digital versatile discs (DVDs). Since they covers almost all important ingredients in modern technology, the REEs are called "The vitamins of modern industry." Thus, it enhances their applications in electronics, manufacturing ceramic pigments, metal alloy, plastic creation, medical field, and energy storage devices.

Among the rare earth metals, Cerium has attracted researcher's attention in recent days due to its remarkable physical and chemical properties [1–5]. Cerium is the most reactive element in the lanthanide series group, when compared with other elements. It is present in different minerals, such as silicates, phosphates, carbonates, monazite, bastnaesite, allanite, and cerite. Out of these various minerals, monazite leads as the principal source of Cerium. Cerium comprises of four isotopes of different percentages, listed as ^{136}Ce (0.2%), ^{138}Ce (0.3%), ^{140}Ce (88.5%), and ^{142}Ce (11.1%) [5]. It occurs more abundantly than lead and tin in the earth's crust. Based on the electronic structure, the energy levels of the inner 4f level and 6 s level are the same, which gives different occupancies of two levels. Compared with other rare earth metals, Cerium exists in two oxidation states of Ce^{3+} and Ce^{4+}, in which Ce^{4+} oxidation mode is more stable than Ce^{3+} due to its [Xe]$4f^{0}$ level of electronic structure. Cerium is a soft, malleable, and ductile rare earth metal. Its color varies from iron gray to silvery nature. When it is scratched with a sharp knife, it starts to ignite, once it is in pure form. Cerium was first applied in gas mantles, invented by the Austrian Chemist Carl Auer Von Welsbach. Mischmetal is a pyrophoric alloy that is composed of 50% of Cerium, which is widely applied in lighter flints. This addition of mischmetal to steel is made for improving the mechanical properties [6]. By the addition of Cerium, the photostability of pigments is enhanced, providing light fastness pigments [7]. In television, F-centre defects are created due to the continuous bombardment of electrons, which may cause a darkening effect on the television glass plates, which is reduced by the addition of Cerium to the cathode ray tube (CRT). Cerium is employed as a dopant for phosphorus in lamps, white light-emitting diodes (LEDs), and cathode-ray tube television (CRT TV) screens [8, 9].

1.1 Properties of cerium

Cerium is a chemical element with the symbol Ce and atomic number 58. It belongs to the lanthanide series of rare earth elements. Cerium exhibits several distinctive properties that are listed in **Table 1**. Applications of Cerium are listed below:

- Cerium finds applications in various industries. It is used as a catalyst in automotive catalytic converters, helping to reduce pollutants in exhaust gases.
- In the glass and ceramics industry, Cerium is employed as a polishing agent.
- Cerium is utilized as an alloying agent in metallurgy, enhancing the properties of alloys.

Cerium	Properties
Chemical symbol	Ce
Atomic number	58
Atomic mass	140.12 g/mol
Density	6.770 g/m^3
Melting point	1065°C
Boiling point	3556°C
Crystal structure	Face Centered Cubic (FCC)
Electrical conductivity	Good electrical conductor
Magnetic property	Paramagnetic at room temperature Ferromagnetic at low temperature
Reactivity	Highly reactive with water, oxygen, and acids

Table 1.
Properties of Cerium.

- It serves as a component in phosphors used in fluorescent lighting and color television tubes.

Electronic properties: Cerium is an excellent electrical conductor, contributing to its applications in certain electronic and catalytic devices.

Natural abundance: Cerium is relatively abundant in the Earth's crust, making up a significant portion of rare earth elements.

Isotopes: Cerium has several isotopes, with the most stable being Cerium-140.

Radioactivity: Some Cerium isotopes are radioactive, posing health and environmental risks. Cerium's diverse properties contribute to its widespread use in various industrial applications, making it a valuable element in technological advancements and material sciences.

Table 2 provides a balanced overview of the advantages and disadvantages associated with the use of Cerium, considering its physical, environmental, and technological aspects.

1.2 Ferrites

Recently, several researchers' works have been done on the fabrication of different ferrites due to their fascinating physical and chemical properties than other metal oxide groups. Ferrites belong to the metal oxide group with a spinel structure of chemical formula MFe_2O_4 (M = Co, Cu, No, Zn, etc.), where M denotes the divalent ions. Ferrites come under the compound of ferrimagnetic ceramics, which is applied in the various opto-electric and magnetic materials due to their notable properties of minimum eddy current losses and high electrical resistances [10–12]. The structure of spinel ferrite is shown in **Figure 2**. In spinel ferrites, MFe_2O_4, the M that represents

Cerium	Advantages	Disadvantages
Physical	• Abundant in Earth's crust • Relatively low cost • Good electrical conductivity • High melting point • Ferromagnetic at low temperatures	• Radioactive isotopes of Cerium pose health and environmental risks • Prone to oxidation in air, requiring careful handling and storage • Limited applications in pure form, often used in alloys or compounds due to reactivity
Environmental	• Used as a catalyst in automotive catalytic converters, reducing pollutants in exhaust gases	• Extraction and processing of Cerium may have environmental impacts • Recycling of Cerium from electronic waste can be challenging • Potential for soil and water contamination in mining operations
Technological	• Phosphors in fluorescent lighting and color TV tubes • Polishing agent in glass and ceramics industry • Alloying agent in metallurgy, enhancing properties of alloys	• Limited applications compared to other rare earth elements

Table 2.
Physical, environmental and technological advantages and disadvantages of Cerium.

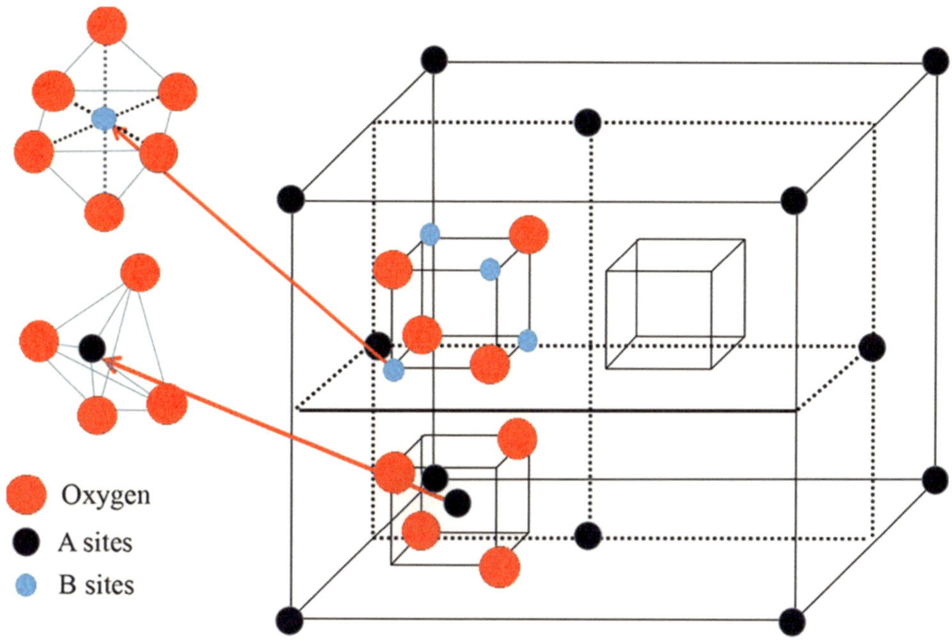

Figure 2.
Structure of spinel ferrite.

the divalent ions and Fe indicates the divalent M^{2+} and trivalent Fe^{3+} ions in the octahedral and tetrahedral sites, the ferrites are classified into normal, inverse, and mixed spinels. In normal spinel, the Fe^{3+} ions are located in the octahedral sites and M^{2+} ions (Cu, Zn, Co, Mg, etc.) occupied in the tetrahedral sites, whereas in inverse spinel, one half of the Fe^{3+} ions occupied in tetrahedral sites and the remaining located in the octahedral sites, and the M^{2+} ions distributed in the octahedral sites. Thus, the magnetic properties of the spinel ferrites depend on the cation distribution between the octahedral and tetrahedral sites [13–16].

Ferrites may be treated as soft and hard ferrites, based on the magnetic property of coercivity (H_c), saturation magnetization (M_s), and retentivity (M_r). Soft ferrites can easily magnetize and demagnetize, the value of coercive field is around less than 40 A/m, which indicates lower value, but high saturation magnetization (M_s) and high permeability (μ_r), whereas for hard ferrites, it is opposite to that of soft ferrites in terms of magnetic properties.

1.3 Impact of Cerium in different ferrites

1.3.1 Cerium-doped cobalt ferrite

Elaya kumar *et al.* [17] reported the X-ray powder diffraction analysis of spinel nanocrystalline $CoCe_xFe_{2-x}O_4$ (x = 0.0, 0.1, 0.2, 0.3, 0.4, and 0.5) samples revealed distinct diffraction patterns which are shown in **Figure 3**, specifically (220), (311), (222), (422), (400), (511), (440), and (533), confirming the formation of single-phase $CoCe_xFe_{2-x}O_4$ with a well-defined cubic structure (JCPDS No. 22–1086). The absence of extra peaks indicates the samples' crystallization in a singular cubic phase. The average crystallite size (D) was calculated using the Scherrer formula for the

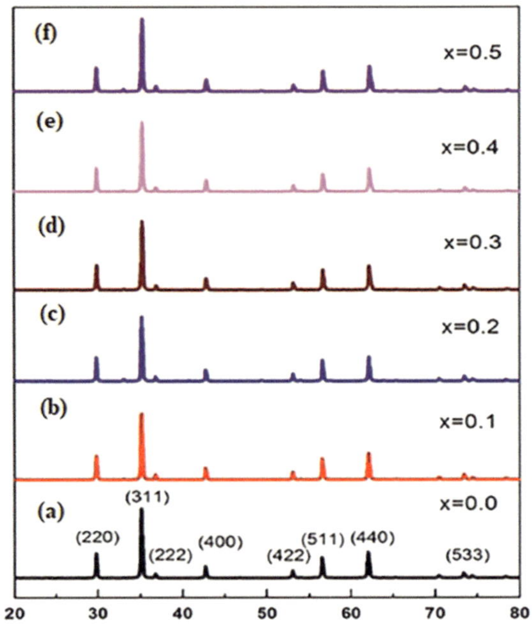

Figure 3.
X-ray diffraction (XRD) pattern of $CoCe_xFe_{2-x}O_4$*(a = x = 0.0, b = x = 0.1, c = x = 0.2, d = x = 0.3, e = x = 0.4, and f = x = 0.5).*

(311) peak, revealing an initial size of 25.36 nm that decreased to a size of 16.53 nm and values are listed in **Table 3**. Magnetic measurements, conducted *via* VSM analysis at room temperature, displayed S-shaped magnetic hysteresis loops that are shown in **Figure 4** for all samples (x = 0.0, 0.1, 0.2, 0.3, 0.4, and 0.5) with non-linear, reversible behavior, observable coercivity, and modest retentivity. Saturation magnetization (Ms), coercive force (HC), and remanence magnetization (Mr) were derived from the M-H loops. Specifically, the saturation magnetization consistently decreased with increasing Cerium content listed in **Table 3**, indicating that Cerium substitution led to a reduction in saturation magnetization of cobalt ferrite, in contrast to previous studies reporting an increase in saturation magnetization on Cerium substitution.

1.3.2 Cerium-doped copper ferrite

The powder XRD patterns of $CuCe_{1-x}Fe_{2-x}O_4$ samples, as depicted in **Figure 5**, align well with the standard values for copper ferrite (JCPDS file: 770010) reported by K. Elaya kumar *et al.* (2019) [18]. The diffraction peaks corresponding to (220), (311), (222), (400), (422), (511), and (440) reflection planes signify the successful attainment of a single-phase cubic structure of copper ferrite through the defined synthesis procedure. This suggests the complete incorporation of Ce^{3+} ions into the spinel lattice of cobalt ferrite, substituting Fe^{3+} ions into the octahedral sites. The larger ionic radius of Ce^{3+} (1.14 Å) compared to that of Fe^{3+} leads to a reduction in lattice parameters and cell volume, as indicated in **Table 3**. The XRD peaks exhibit a gradual shift towards higher values with an increase in Ce^{3+} concentration in the $CuCe_{1-x}Fe_{2-x}O_4$ lattice. The (311) plane was used to evaluate the crystallite size of $CuCe_{1-x}Fe_{2-x}O_4$ samples using the Scherrer formula, revealing a decrease from 25.36 to 18.53 nm for $CuCe_{1-x}Fe_{2-x}O_4$, with x = 0.0, 0.1, 0.2, 0.3, 0.4, and 0.5 samples.

Sample	x	Synthesized method	Structure	Crystallite size (nm)	Magnetic parameters			Reference
					H_C (O_e)	M_r (emu/g)	M_s (emu/g)	
$CoCe_xFe_{2-x}O_4$	0	Sol-gel	Single-phase cubic structure	25.36	106.52	11.22	55.64	[17]
	0.1			22.51	99.48	10.55	52.85	
	0.2			20.98	95.47	9.94	47.87	
	0.3			19.85	93.92	8.91	38.35	
	0.4			18.67	91.95	7.45	35.46	
	0.5			16.53	89.48	5.98	29.32	
$CuCe_{1-x}Fe_{2-x}O_4$	0	Sol-gel	Single-phase cubic structure	25.36	25.45	9.25	4.668	[18]
	0.1			23.88	24.98	8.36	4.548	
	0.2			21.98	23.86	7.63	3.856	
	0.3			20.65	22.68	6.85	3.365	
	0.4			19.39	21.83	5.43	2.964	
	0.5			18.53	20.75	4.82	2.322	
$CuFe_{2-x}Ce_xO_4$		Auto-combustion sol-gel	Tetragonal phase structure	52	1200	9	15	[19]
$NiCe_xFe_{2-x}O_4$	0	Sol-gel	Single-phase cubic structure	23.5	0.0289	0.358	65.38	[20]
	0.1			22.8	0.0274	0.312	47.53	
	0.2			22.3	0.0192	0.296	37.34	
	0.3			21.7	0.0189	0.281	24.32	
	0.4			21.2	0.0176	0.277	21.22	
	0.5			20.9	0.0169	0.265	8.685	

Sample	x	Synthesized method	Structure	Crystallite size (nm)	Magnetic parameters			Reference
					H_C (O_e)	M_r (emu/g)	M_s (emu/g)	
$NiCe_xFe_{2-x}O_4$	0	Chemical co-precipitation method	Pure spinel phase structure	55.8	63.5	7.27	51.9	[21]
	0.05			42.2	92.2	7.17	46.9	
	0.10			39.2	118.6	5.05	36.5	
	0.15			33.5	130.9	3.19	22.6	
	0.20			41.8	50.2	2.40	25.8	
$NiCe_xFe_{2-x}O_4$	0	Chemical route method	Pure spinel phase structure	57	41.84	—	61.6	[22]
	0.01			28	42.90	—	67.7	
	0.02			—	41.34	—	25.16	
	0.03			—	38.67	—	28.4	
	0.04			11	41.14	—	14.46	
	0.06			19	40.78	—	5.9	
	0.08			—	41.35	—	15.1	
	0.10			—	31.64	—	26.83	
$MgCe_xFe_{2-x}O_4$	0.01	Sol-gel method	Face-centred cubic structure	51.113	140.86	0.13650	0.51660	[23]
	0.012			82.12	134.46	0.14146	0.64289	
	0.014			46.61	94.586	0.0775	0.65246	
	0.016			67.22	144.40	0.18184	0.83079	
$MgCe_xFe_{2-x}O_4$	0	Sol–gel auto-combustion method	Single-phase spinel structure	45.26	120	2.08	13.59	[24]
	0.02			42.42	66.97	0.39	7.40	
	0.04			42.82	95.26	1.11	9.65	
	0.06			46.47	112.33	2.61	17.42	

DOI: http://dx.doi.org/10.5772/intechopen.1004167

Sample	x	Synthesized method	Structure	Crystallite size (nm)	Magnetic parameters			Reference
					H_C (O_e)	M_r (emu/g)	M_s (emu/g)	
$SrFe_{12-x}Ce_xO_{19}$	0	Sol-gel auto-combustion method	—	77	—	—	68.56	[25]
	0.1			67	—	—	58.45	
	0.2			65.1	—	—	53.12	
	0.3			47.7	—	—	51.24	
	0.4			63.8	—	—	47.41	
	0.5			51.1	—	—	46.69	

Table 3.
Synthesize method, structure, crystallite size and magnetic parameters of Cerium-doped different nanoferrites.

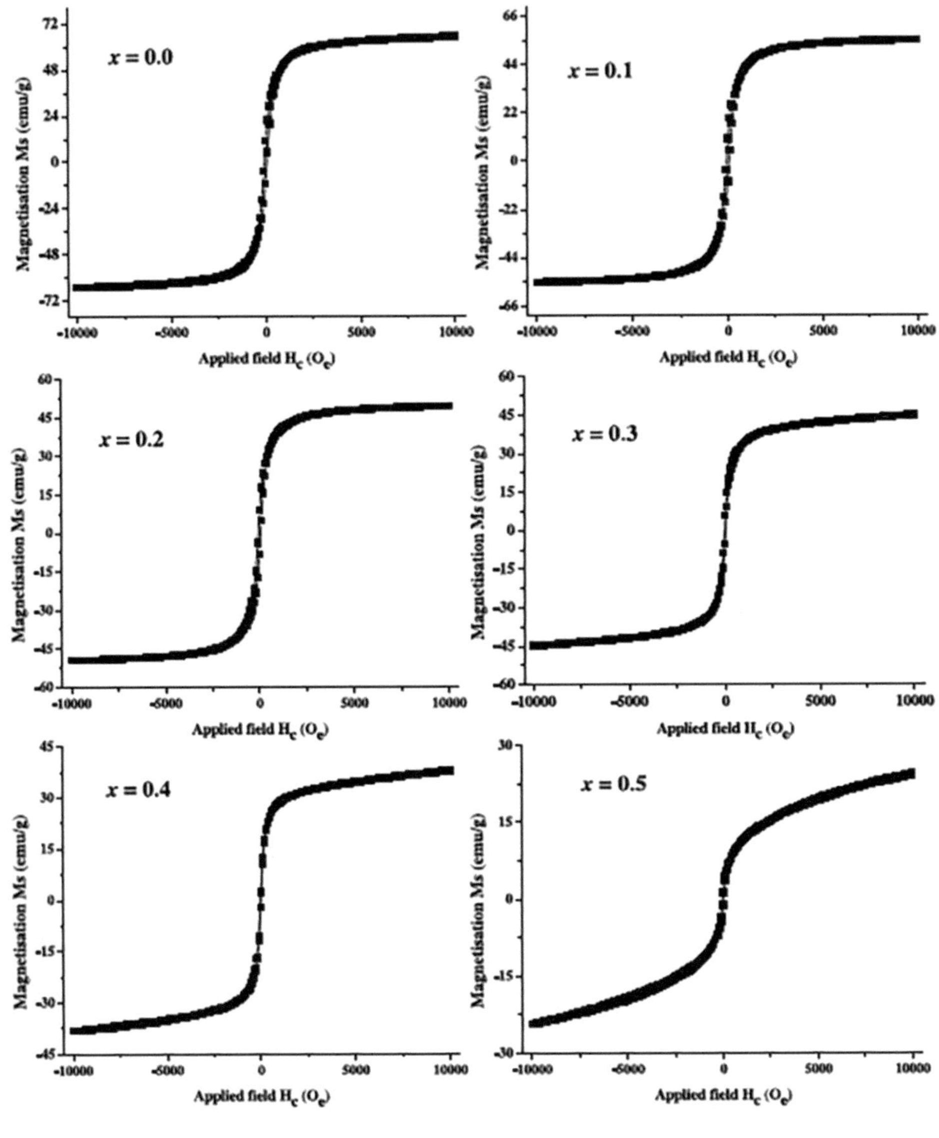

Figure 4.
Vibrating sample magnetometer (VSM) analysis of $CoCe_xFe_{2-x}O_4$*(a = x = 0.0, b = x = 0.1, c = x = 0.2, d = x = 0.3, e = x = 0.4, and f = x = 0.5).*

The VSM analysis at room temperature recorded magnetic measurements of spinel $CuCe_{1-x}Fe_{2-x}O_4$ (x = 0.0, 0.1, 0.2, 0.3, 0.4, and 0.5) samples, and the magnetic hysteresis (M-H) loops in **Figure 6** display well-defined S-shaped curves. Saturation magnetization (M_s), coercivity (H_c), and remanence magnetization (M_r) were calculated from the M-H loop. The substitution of Ce^{3+} ions into the copper ferrite matrix significantly altered its magnetic properties, as shown in **Table 3**. Saturation magnetization (Ms) should decrease from 4.668 (emu/g) at lower Cerium dopant concentrations to 2.322 (emu/g) at higher Cerium dopant concentrations. The reduction in Ms. can be attributed to the substitution of Ce^{3+} ions for Fe^{3+} ions at the B sites, resulting in a decrease in the magnetic moments of the unit cells. The

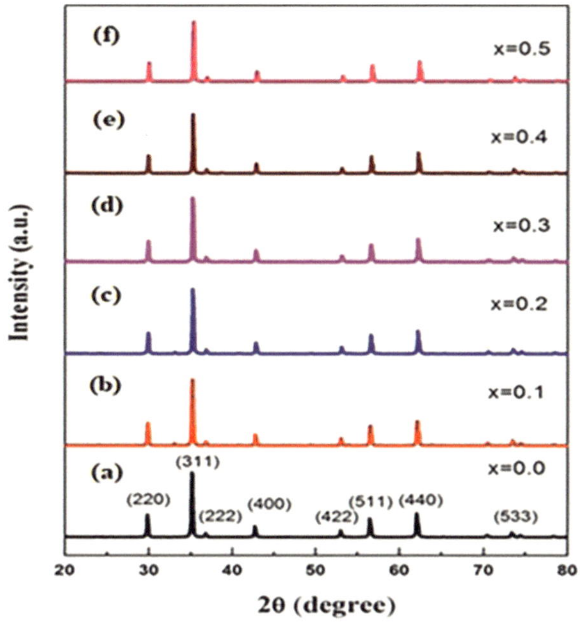

Figure 5.
X-ray diffraction (XRD) spectra of Ce^{3+} doped copper ferrite (a = x = 0.0, b = x = 0.1, c = x = 0.2, d = x = 0.3, e = x = 0.4, and f = x = 0.5).

observed reduction in Ms. suggests an increase in Ms. upon Cerium substitution into copper ferrite.

Tetragonal phase of spinel copper ferrite nanoparticles is confirmed by the XRD pattern, which was reported by Mehdi Rahimi-Nasrabadi et al. [19]. The average crystalline size of $CuFe_2O_4$ nanoparticles is found to be 52 nm by the auto-combustion sol-gel method. Furthermore, $CuFe_{2-x}Ce_xO_4$ nanoparticles exhibit the ferromagnetic characteristics at room temperature. From the hysteresis loop, the saturation magnetization (M_s) value is said to be 15 emu/g, the remnant magnetization (M_r) is noted as 9 emu/g, and the coercive field (H_c) value is 1200 Oe, respectively, and these values are listed in **Table 3**.

1.3.3 Cerium-doped nickel ferrite

The structural study of Cerium-doped Nickel ferrite nanoparticles, $NiCe_xFe_{2-x}O_4$, by the sol-gel method was successfully evaluated through the XRD technique, as reported by Elayakumar et al. [20]. The XRD patterns reveal that the pure-phase formation of spinel structural crystalline samples is shown in **Figure 7**. It shows that the Cerium ions (Ce^{3+}) are entered into the Nickel ferrite lattice without any distortion. With the help of Scherrer formula, the crystalline sizes of the resultant samples are calculated and the values are listed in **Table 3**. The ionic radius of Ce_{3+} is 1.03 Å, which is greater than the ionic radius of Fe^{3+} that is 0.64 Å, which results in the impact of Cerium ions to penetrate into the spinel lattice. It was found that the crystalline size reduced from 23.5 to 20.9 nm, when the Cerium ions' concentration was increased. The magnetic characterization of $NiCe_xFe_{2-x}O_4$ nanoparticles was analyzed by the vibrating sample magnetometer (VSM), which is shown in **Figure 8**. The magnetic parameters like saturation magnetization (M_s), coercivity (H_c), and remanence

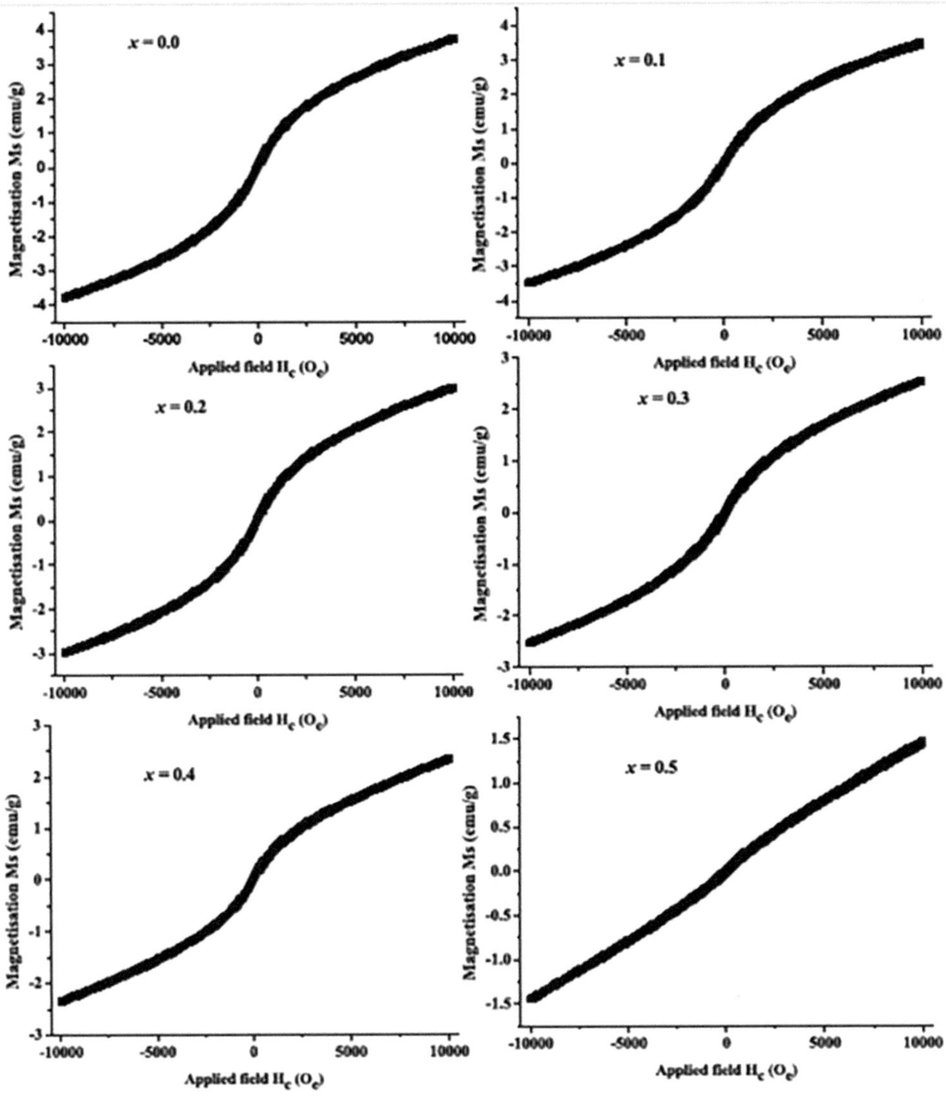

Figure 6.
Vibrating sample magnetometer (VSM) analysis of Ce^{3+} doped copper ferrite (a = x = 0.0, b = x = 0.1, c = x = 0.2, d = x = 0.3, e = x = 0.4, and f = x = 0.5).

magnetization (M_r) were calculated. It was noted that the remarkable changes in the magnetic parameters of Nickel ferrite lattice took place due to the replacement of Fe^{3+} ions by Ce^{3+}. Thus, it is weakening the sublattice interaction and creates reduction in the coercivity value of the products. The remanence magnetization (M_r) value was decreased to 0.265 from 0.365 emu/g and value are listed in **Table 3**, which indicates that the prepared samples exhibit superparamagnetic nature. Thus, it happens due to the Ce^{3+}-Fe^{3+} interaction taking place on account of 3d-4f electron coupling.

Priyadharshini et al. [21] synthesized Ce-doped $NiFe_2O_4$ nanoparticles by chemical co-precipitation method. It was noted that the lattice parameters are altered by the substitution of Cerium-doped Nickel ferrite lattice. By the substitution of Scherrer method, the crystalline size was gradually reduced from 55.8 to 41.8 mm

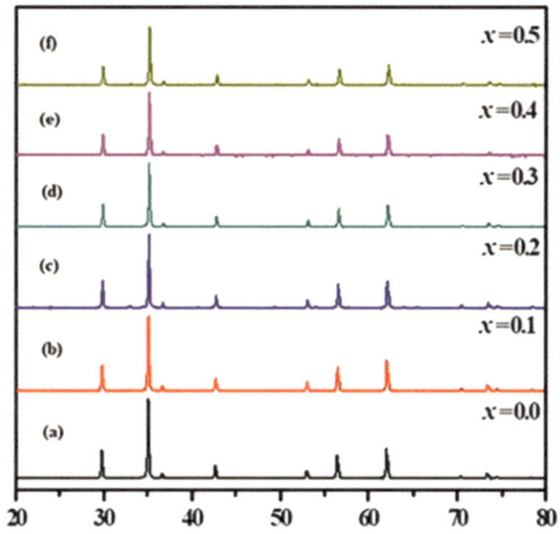

Figure 7
Powder XRD patterns of $NiCe_xFe_{2-x}O_4$ (a = x = 0.0, *b = x = 0.1, c = x = 0.2, d = x = 0.3, e = x = 0.4, and f = x = 0.5) samples.*

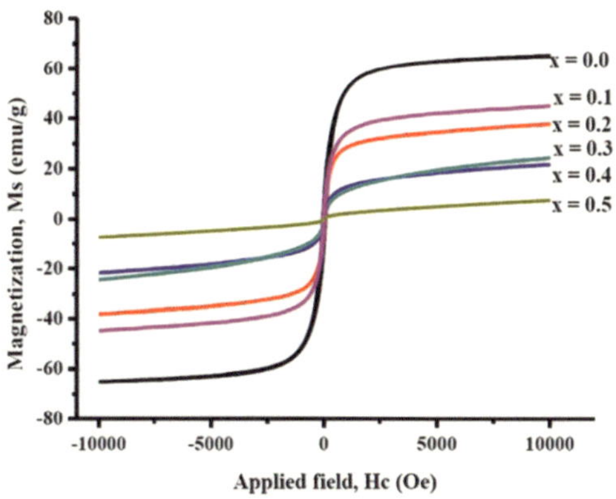

Figure 8.
Vibrating sample magnetometer (VSM) analysis of $NiCe_xFe_{2-x}O_4$ (x = 0.0, 0.1, 0.2, 0.3, 0.4, and 0.5) samples.

with the substitution of Ce^{3+} ions and is listed in **Table 3**. When the amount of Cerium dopant increases, the $NiFe_2O_4$ lattice limits the crystalline growth. Well-deserved 'S'-shaped curves are obtained from the VSM graph. The saturation magnetization (M_s), remanence value (M_r), and coercivity (H_c) values are altered with the dopant concentration of Ce^{3+} ions. The coercivity value (H_c) slowly increases from 63.05 Or to 130.9 Oe at the concentration of (x = 0.15 M) Cerium ions penetrating into the spinel lattice. The Ms. and Mr. values reduced to 22.6 and 3.19 emu/g when the Ce content is increased at (x = 0.15 M). Thus, it happens due to the weakening of the Fe^{3+}-Fe^{3+} interactions, brought about by the replacement of Fe^{3+} ions by the Ce^{3+} ions.

Gagan Dixit et al. [22] synthesized the Ce-doped $NiFe_2O_4$ nanoparticles and characterized the structural, optical, and magnetic properties by the chemical route method. XRD reveals the pure-phase formation of spinel structure nanoparticles. The substitution of Ce^{3+} ions into the spinel lattice creates lattice strain and thus reduces the crystalline size from 57 to 17 nm, which is listed in **Table 3**. However, the other structural parameters remain constant. The magnetic parameters like saturation magnetization (M_s) and coercivity (H_c) are calculated from the VSM graph. The Ms. value is gradually decreased to 31.64 from 41.84 emu/g at the dopant concentration level of x = 0.10 M, and the Cerium dopant level is listed in **Table 3**. Thus, the Ce^{3+} ions penetrate the Nickel ferrite lattice.

1.3.4 Cerium-doped magnesium ferrite

Muthuraman et al. [23] reported the polycrystalline and single-phase formation of prepared samples, which was confirmed by the XRD techniques. The structural and magnetic properties of Cerium-doped Magnesium ferrite are studied by the XRD and VSM techniques. These Cerium-doped Magnesium ferrite nanoparticles are prepared by the sol-gel method. XRD graphs show the impurities-free Cerium-doped Magnesium ferrite nanoparticles. The structural parameters like lattice parameter and crystallite size were calculated from Bragg's law and Debye Scherrer's formula. The average crystalline size found to be in the range of 46.61 to 82.12 nm values is listed in **Table 3**. The lattice parameter increases from 8.3732 to 8.4076 at the Cerium concentration range of x = 0.014 M. From the hysteresis loop, magnetic parameters like saturation magnetization (Ms), coercivity (Hc), and retentivity (Mr) are measured. The least saturation magnetization value of 0.51660 emu was noted at a lower concentration of Ce^{3+} dopant and gradually increases to 0.83079 emu at the higher concentration of Cerium ions. The lower coercivity value of 95.586 Oe was found at the concentration of x = 0.016 M. And it was found to have the least retentivity value of 0.0775 at the concentration of x = 0.016 M values that are listed in **Table 3**.

Syed Ismail Ahmad et al. [24] reported the Cerium-doped Magnesium ferrite nanoparticles by the sol-gel auto-combustion technique. X-ray diffraction analysis reveals the single-phase spinel structure without any impurities. The addition of Ce^{3+} ions in the parental lattice decreases the intensity of the peaks, which indicates the strain produced due to the creation of oxygen vacancies. Thus, the reflections are gradually shifted to lower angles because the radii of Ce^{3+} ions are larger than those of the Fe^{3+} ions. The average crystallite size in the range of 42–46 nm values is listed in **Table 3**. The magnetic parameters were calculated from the M-H curves of vibrating sample magnetometer (VSM) device for the Cerium-doped Magnesium ferrite. Due to the porosity of the samples, the saturation magnetization (M_s) value decreases from 13.59 to 9.65 emu/g at the concentration range of Cerium (x = 0.04). After this saturation, magnetization slowly increases to 17.42 emu/g at the dopant range of (x = 0.06) concentration. Thus, the increase in (M_s) value is attributed to the migration of Mg^{2+} ions from B site to A site and segregation of Ce^{3+} ions at the level of grain boundary. The large amount of Ce^{3+} ion dopant may cause redistribution of cations, which means the divalent magnesium ions are partially placed in A site and equal number of Fe^{3+} ions move to B site, which makes the increase in saturation magnetization (M_s). The coercivity value (H_c) was found to be 120 Oe at (x = 0.00) of Cerium content and decreases to 95.26 Oe at (x = 0.04) of Cerium content. Once again, the coercivity value (H_c) increases to 112.33 Oe at higher concentration of (x = 0.06) of Cerium dopant into the Magnesium Ferrite values and is listed in **Table 3**. Overall, the coercivity (H_c) value decreases when compared with undoped Magnesium ferrite, since the coercivity (H_c) purely depends on the crystal anisotropy. From the magnetic parameters, it was concluded that all the samples exhibit superparamagnetic nature.

1.3.5 Cerium-doped strontium hexaferrite

$SrFe_{12-x}Ce_xO_{19}$ magnetic nanoparticles are fabricated by the sol-gel auto-combustion method. The structural parameters were evaluated by the XRD technique at room temperature. All the noted peaks are well matched with the JCPDS Card No. 96–100-8857, which indicates the impurities-free Cerium-doped Strontium Hexaferrite. The ionic radius of Ce^{3+} is 1.05, which is larger than that of Fe^{3+} which is 0.64, and thus the replacement of Fe^{3+} ion by the Ce^{3+} ion takes place. The concentration of crystal axis takes place due to the strain, defects, and chemical composition of Cerium-doped Strontium Hexaferrite. The magnetic measurements were noted from the vibrating sample magnetometer technique. The saturation magnetization (M_s) value is 68.56 emu/g at 300 k that was noted and is listed in **Table 1**. The coercive field (H_c) of the Strontium Hexaferrite was 4337–5703 Oe at room temperature and 2808–4514 Oe at 10 k. The remanence magnetization (M_r) is measured in the range of 35–23 emu/g at room temperature and 50–32 emu/g at 10 k. These overall magnetic parameters reveal that the formation of Cerium-doped Strontium Hexaferrite exhibits the ferromagnetic characteristics. The magnetization expression is given by Eq. (1)

$$M = M_s\left[1 - \frac{\alpha}{H} - \frac{\beta}{H^2}\right] + \chi H \quad (1)$$

where χ – high field susceptibility.

The values of saturation magnetization (M_s) are higher at lower temperature, when compared with those at room temperature. This may happen due to the reduction in thermal fluctuations of magnetic moments and thus strengthening the superexchange interaction.

2. Conclusion

Researchers have been drawn to Cerium-doped magnetic ferrites, a rare earth element, because of its distinct characteristics and wide range of uses in several sectors. The chapter provides a summary of the structural characteristics, including crystallite size and structures. Saturation magnetization (M-S) and coercivity values for several ferrites are listed, and other magnetic properties are recorded from the VSM investigation. Sol-gel analysis reveals that the majority of Cerium-doped magnetic ferrites have a single-phase spinel structure. Among these several ferrites doped with Ce^{3+} ions, the nanoparticles of $CuFe_2O_4$ and $MgFe_2O_4$ exhibit the least saturation magnetization (Ms). The crystalline size of all ferrites synthesized using the sol-gel technique is around 50 nm on average. As a result, the Cerium ions are successfully revealed by the differences in crystallite size and magnetic properties in all magnetic ferrites.

Author details

Kannan Elaya Kumar[1*], S. Muthulingam[1] and A. Manikandan[2]

1 St. Joseph's College of Arts and Science (Autonomous), Cuddalore, Tamilnadu, India

2 Karpagam Academy of Higher Education, Coimbatore, Tamil Nadu, India

*Address all correspondence to: elaya.phys@gmail.com

References

[1] Lide DR, editor. CRC Handbook of Chemistry and Physics. Boca Raton, FL: CRC press; 2004

[2] Mogensen M, Sammes NM, Tompsett GA. Physical, chemical and electrochemical properties of pure and doped ceria. Solid State Ionics. 2000;**129**(1-4):63-94

[3] Kilbourn BT. Cerium and Cerium compounds. In: Kirk-Othmer Encyclopedia of Chemical Technology. Vol. 1-23. Wiley Online Library; 2000

[4] Haxel G. Rare Earth Elements: Critical Resources for High Technology. US: US Department of the Interior, US Geological Survey; 2002

[5] O'Neil MJ. The Merck index: And encyclopedia of chemicals, drugs, and biologicals. In: The Merck Index: And Encyclopedia of Chemicals, Drugs, and Biologicals. 2006, 1756-1756

[6] Greenwood NN, Earnshaw A. Chemistry of the Elements. Burlington: Elsevier; 2012

[7] Gleń M, Grzmil B, Sreńscek-Nazzal J, Kic B. Effect of CeO_2 and $Sb2O_3$ on the phase transformation and optical properties of photostable titanium dioxide. Chemical Papers. 2011;**65**(2):203-212

[8] DaNa - Information on the safety of new, innovative materials and nanomaterials. Cerium Dioxide, 2024. Available from: https://www.mcgill.ca/crcf/files/crcf/2005-Increasing_Caseloads_OIS93-98.pdf

[9] Trovarelli A. Catalysis by Ceria and Related Materials. London: World Scientific; 2002

[10] Rashad MM, Rayan DA, Turky AO, Hessien MM. Effect of Co2+ and Y3+ ions insertion on the microstructure development and magnetic properties of Ni0. 5Zn0. $5Fe_2O_4$ powders synthesized using co-precipitation method. Journal of Magnetism and Magnetic Materials. 2015;**374**:359-366

[11] Mund HS, Ahuja BL. Structural and magnetic properties of Mg doped cobalt ferrite nano particles prepared by sol-gel method. Materials Research Bulletin. 2017;**85**:228-233

[12] Moradmard H, Shayesteh SF, Tohidi P, Abbas Z, Khaleghi M. Structural, magnetic and dielectric properties of magnesium doped nickel ferrite nanoparticles. Journal of Alloys and Compounds. 2015;**650**:116-122

[13] Xu J, Yang H, Fu W, Du K, Sui Y, Chen J, et al. Preparation and magnetic properties of magnetite nanoparticles by sol–gel method. Journal of Magnetism and magnetic Materials. 2007;**309**(2):307-311

[14] Vinayak V, Khirade PP, Birajdar SD, Gaikwad PK, Shinde ND, Jadhav KM. Low temperature synthesis of magnesium doped cobalt ferrite nanoparticles and their structural properties. International Advanced Research Journal in Science Engineering and Technology. 2015;**2**(3):55-58

[15] Kumar R, Kumar H, Singh RR, Barman PB. Variation in magnetic and structural properties of Co-doped Ni–Zn ferrite nanoparticles: A different aspect. Journal of Sol-Gel Science and Technology. 2016;**78**:566-575

[16] Kharisov BI, Dias HR, Kharissova OV. Mini-review: Ferrite nanoparticles in the

catalysis. Arabian Journal of Chemistry. 2019;**12**(7):1234-1246

[17] Elayakumar K, Dinesh A, Manikandan A, Palanivelu M, Kavitha G, Prakash S, et al. Structural, morphological, enhanced magnetic properties and antibacterial bio-medical activity of rare earth element (REE) Cerium (Ce^{3+}) doped $CoFe_2O_4$ nanoparticles. Journal of Magnetism and Magnetic Materials. 2019;**476**:157-165

[18] Elayakumar K, Manikandan A, Dinesh A, Thanrasu K, Raja KK, Kumar RT, et al. Enhanced magnetic property and antibacterial biomedical activity of Ce^{3+} doped $CuFe_2O_4$ spinel nanoparticles synthesized by sol-gel method. Journal of Magnetism and Magnetic Materials. 2019;**478**:140-147

[19] Rahimi-Nasrabadi M, Behpour M, Sobhani-Nasab A, Jeddy MR. Nanocrystalline Ce-doped copper ferrite: Synthesis, characterization, and its photocatalyst application. Journal of Materials Science: Materials in Electronics. 2016;**27**:11691-11697

[20] Elayakumar K, Sathana V, Kumar RT. Structural and magnetic characterization of rare earth element Cerium-doped nickel ferrite nanoparticles ($NiCe$ x Fe_2-x O_4) by sol-gel method with antibacterial activity. Journal of Superconductivity and Novel Magnetism. 2020;**33**(7):2171-2178

[21] Priyadharshini P, Pushpanathan K. Synthesis of Ce-doped $NiFe_2O_4$ nanoparticles and their structural, optical, and magnetic properties. Chemical Physics Impact. 2023;**6**:100201

[22] Dixit G, Singh JP, Srivastava RC, Agrawal HM. Structural, optical and magnetic studies of Ce doped $NiFe_2O_4$ nanoparticles. Journal of Magnetism and Magnetic Materials. 2013;**345**:65-71

[23] Muthuraman K, Naidu V, Ahmed SK, Vasudevan T. Study of electrical and magnetic properties of Cerium doped nano smart magnesium ferrite material. International Journal of Computer Applications. 2013;**65**(23):18-24

[24] Ahmad SI, Kumar DR, Syed IA, Satar R, Ansari SA. Structural, spectroscopic and magnetic study of nanocrystalline Cerium-substituted magnesium ferrites. Arabian Journal for Science and Engineering. 2017;**42**:389-398

[25] Almessiere MA, Slimani Y, Baykal AJ. Structural and magnetic properties of Ce-doped strontium hexaferrite. Ceramics International. 2018;**44**(8):9000-9008

Chapter 5

Innovative Applications of Cerium Oxide-Based Materials in Civil Engineering, Automation, and Energy Sectors

Stephan Kozhukharov, Christian Girginov, Vanya Lilova and Plamen Petkov

Abstract

Cerium oxide materials exhibit remarkable properties, positioning them as highly effective, environmentally friendly solutions across diverse applications. This chapter provides a comprehensive overview of fundamental concepts and technological methodologies related to cerium oxide (CeO_2) and doped ceria-based materials. Emphasis is placed on electrochemical deposition, spray pyrolysis, and the sol-gel approach for synthesizing thin and thick layers of ceria. The versatility of these materials is explored, spanning from corrosion protection layers and specialized ceramic elements for sensor applications to components for solid oxide fuel cells (SOFCs) and electrodes for water-splitting cells. Additionally, the chapter delves into the promising applications of recently developed ceria-based nanomaterials in various fields, marking some advanced methods for CeO_2-based materials synthesis. The key findings are succinctly summarized in the concluding section.

Keywords: corrosion protection, cerium conversion coatings, sensor components, solid oxide fuel cells, water-splitting cells

1. Introduction

In recent times, there has been a growing interest in lanthanides, particularly Cerium (Ce) [1]. This surge is attributed to their economic competitiveness [2], with Cerium, in particular, being as abundant as copper [3]. Moreover, the Rare Earth (RE) elements are recognized for their low toxicity, posing no harm to health through ingestion or inhalation [4, 5]. Consequently, various Cerium compounds are now being explored for prospective applications in diverse sectors, such as electrocatalytic hydrogen generation [6–8], thermochemical fuel production [9–11], water-splitting devices [12–20], biodiesel production catalysts [21, 22], sensors [23–28], biosensors [29–32], and more. Recent investigations into the use of ceria nanoparticles for medical applications [33–41] underscore the potential of cerium oxide-based nanomaterials.

2. Application of Cerium oxide and soluble salts-based materials for corrosion protection

Highly doped metal alloys serve as essential construction materials for various purposes [42, 43]. Like any solid-state industrial product, their exceptional mechanical strength is derived from both their chemical composition and structure (**Figure 1**).

The strength enhancement of these materials is primarily influenced by intermetallic inclusions [44] and the granulometry of the basic metallic lattice [45, 46]. However, the presence of doping elements leads to galvanic corrosion [47–49], which can progress into other localized corrosion forms [50], such as crystallographic pitting, intergranular corrosion [51], and crevice corrosion [52, 53]. These corrosion types can lead to stress corrosion cracking [47, 50, 54]. Another factor contributing to the increased vulnerability of highly doped alloys compared to low-doped ones is the disruption of the native oxide layer on their surface [55]. Additionally, the native oxide layers on Al-based alloys demonstrate amphoteric behavior, rendering them susceptible to dissolution in both alkaline and acidic environments [50, 56, 57].

The foundational principles governing the use of cerium compounds for efficient metal corrosion mitigation through inhibition and the formation of conversion coatings were first proposed by the groundbreaking works of Hinton et al. [58, 59] in the mid-1980s, followed by Davenport et al. [60, 61] and Mansfeld et al. [62, 63] in the early 1990s. Since then, the interest in cerium conversion coatings has steadily grown, particularly in optimizing the conditions for their deposition.

The initial research on lanthanides focused on their cathodic inhibition ability, arising from the reaction between lanthanide ions and hydroxyl ions produced by the oxygen reduction reaction (ORR), as depicted in Eqs. (1)–(5) [64–66]:

$$O_2 + 2H_2O + 4e^- \rightarrow 4OH^- \; E° = 0.401 \; V_{SHE} \quad (1)$$

$$O_2 + 2H_2O + 4e^- \rightarrow H_2O_2 + 2OH^- \; E° = 0.682 \; V_{SHE} \quad (2)$$

$$H_2O_2 + 2e^- \rightarrow 2OH^- \; E° = 0.710 \; V_{SHE} \quad (3)$$

$$H_2O_2 + 2e- \rightarrow 2H_2O \; E° = 1.776 \; V_{SHE} \quad (4)$$

$$H_2O_2 + 2H_3O^+ + 2e^- \rightarrow 2H_2O \; E° = 1.776 \; V_{SHE} \quad (5)$$

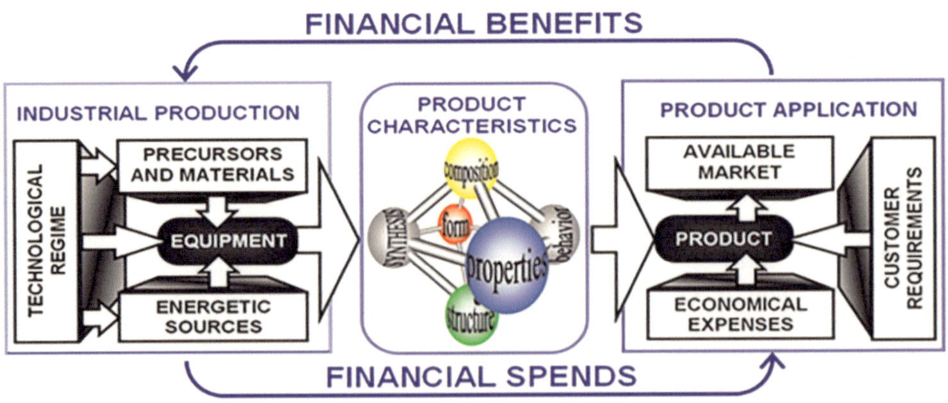

Figure 1.
Correlations between the chemical composition, structure, form, and properties that predetermine the application of any solid-state product.

The hydroxyl anions/OH^- anions generated by these reactions react with dissolved Ce^{3+} and Ce^{4+} ions, forming insoluble cerium hydroxide precipitates according to Eqs. (6) and (7):

$$Ce^{3+} + 3OH^- \rightarrow Ce(OH)_3 \tag{6}$$

$$Ce^{4+} + 4OH^- \rightarrow Ce(OH)_4 \tag{7}$$

These reactions lead to the sealing of cathodic areas on the AA2024 alloy through the deposition of insoluble precipitates, as illustrated in **Figure 2**.

In this context, Bethencourt et al. conducted a comparative study on the corrosion inhibition efficiencies of chlorides from various lanthanide elements [68]. As a primary outcome, the authors determined that $CeCl_3$ exhibited superior efficiency compared to other chlorides, including elements such as Pr, Nd, Ni, Co, La, Y, and Fe.

Recent elucidations on the inhibitor efficiency of various cerium salts, including inorganic ones such as $CeCl_3$ [69–74], $Ce(NO_3)_3$ [74–80], $Ce(SO_3)_2$ [81], $Ce_2(MoO_4)_3$ [82], as well as some organic Ce-salts like $(CH_3COO)_3Ce$ [83, 84], $(HOC_2H_5COO)_3Ce$ [85], and other Ce-carboxylates [86], have been reported. Furthermore, the substantial potential of Ce-compounds as corrosion inhibitors is highlighted in various review papers [87–93].

Subsequently, the apparent cathodic inhibition capability of cerium-based compounds found successful application in the deposition of cerium conversion coatings (CeCC). The possibility of forming continuous coating layers through CeCC deposition was realized by Conde et al. [94], which was confirmed by other authors [95]. These researchers proposed additional chemisorption processes on the superficial oxide layer of aluminum, enabling the complete coverage by a uniform Ce-O-Al layer. Consequently, CeCC formation initiates from nobler intermetallic inclusions, sealing them completely, and subsequently spreading over the remaining surface. Thus, CeCC growth undergoes a transformation in the formation mechanism from island-like nucleation to subsequent layer formation [96], as illustrated in **Figure 3**.

However, as established by Rodriguez et al. [97], the practical deposition of CeCC faces some challenges related to undesirable precipitation rather than the formation of uniform, dense, and adhesive layers. Recent evidence suggests that the optimal concentration of the Ce-precursor should range between 0.03 and 0.05 mol dm^{-3} [98, 99], and the maximum amount of added 30% H_2O_2 as a deposition activator should be around 10 ml for up to 1000 ml of precursor solution. Experiences thus far

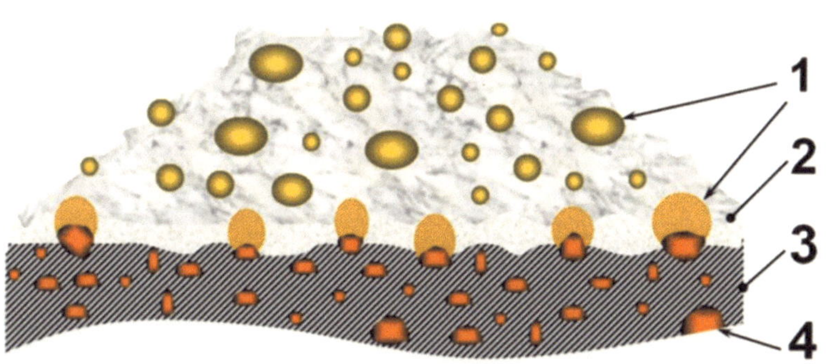

Figure 2.
Visual model of Al-Cu alloy with Ce-inhibitor deposits on its surface: (1) Inhibitor deposits; (2) Surface oxide layer; (3) Aluminum matrix; (4) Cathodic copper intermetallic inclusions [67].

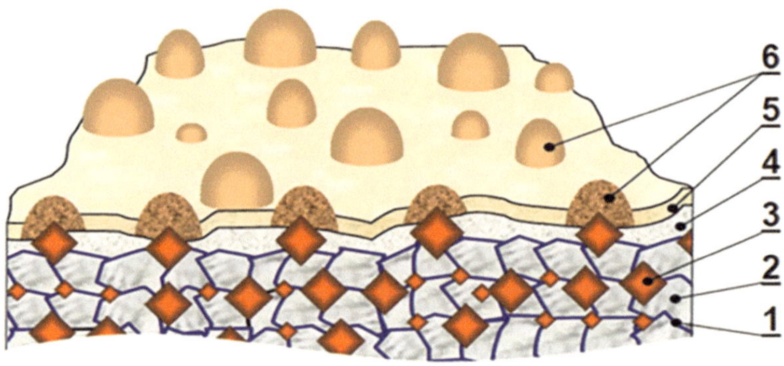

Figure 3.
Illustration of CeCC layer on Cu-doped Al alloy: (1) Grain boundaries; (2) Grains of the basic Aluminum matrix; (3) Cathodic copper intermetallic inclusions; (4) Native oxide layer; (5) CeCC layer; (6) Initial Ce-oxide/hydroxide deposits [96].

indicate that the deposition of these coating layers can be accomplished through dip-coating [96, 97], electrodeposition under galvanostatic [46] or potentiostatic [100, 101] regimes. In all cases, CeCC formation requires activation energy in the form of elevated temperatures (up to 60°C to avoid considerable thermal H_2O_2 decomposition), applied potential (up to −1 V) vs. Ag/AgCl reference electrode, or defined current density (from −2 to −5 mA cm^{-2}). Superior values of these parameters result in $Ce(OH)_3/Ce(OH)_4$ precipitation rather than thin, uniform CeCC formation.

X-ray photoelectron spectroscopy (XPS) studies [102, 103] have revealed that CeCC layers, deposited from Ce(III) soluble salts, convert to Ce(IV) oxides and hydroxides, primarily CeO_2. This conversion occurs due to the more complex mechanism of cerium oxide and hydroxide formation, proceeding under the given regime via the following reactions:

$$4Ce^{3+}(aq) + O_2(aq) + 4OH^- + 2H_2O \rightarrow 4Ce(OH)_2^{\ 2+}(aq) \tag{8}$$

$$2Ce^{3+}(aq) + H_2O_2(aq) + 2OH^-(aq) \rightarrow 2Ce(OH)_2^{\ 2+}(aq) \tag{9}$$

The Ce(IV) ions subsequently precipitate as hydroxides (reaction (10)) and oxides (reaction (11)).

$$2Ce(OH)_2^{\ 2+}(aq) + 2OH^-(aq) \rightarrow Ce(OH)_4(s) \tag{10}$$

$$Ce(OH)_4(s) \rightarrow CeO_2(s) + 2H_2O \tag{11}$$

Additionally, Covelo et al. [104] propose direct oxidation, as indicated by the reaction below (Eq. (12)):

$$4Ce^{3+} + 12OH^- + O_2 \rightarrow 2CeO_2 + 6H_2O \tag{12}$$

Furthermore, the oxidation of Ce^{3+} ions to Ce(IV)-oxides and hydroxides could be further enhanced under conditions of electrochemical deposition, following reaction (13):

$$Ce^{4+} \rightarrow Ce^{3+} + e^- \ E^o = 1.61\ V_{SHE} \tag{13}$$

The deposition of uniform, dense, and adherent CeCC films, instead of obtaining discrete Ce-precipitates, poses a significant challenge, primarily due to the complicated chemical mechanism. Beyond the influence of Ce-compound type and concentration, and, of course, the H_2O_2 activator content, the deposition mechanism and rate can be influenced by regulating various factors, such as pH, temperature, and the presence of additives, among others. Various additives have been shown to have a beneficial effect, including black cuprous oxide "smut" [105], Al^{3+} and Cl^- ions [106], pH buffers [107], etc. Similar studies are also dedicated to the formation of CeCC layers on other metallic substrates, such as galvanized [108] and stainless [109] steel, and zinc substrates [110]. The works of Mansfeld et al. on Ce-Mo [62], Jiang et al. [111] on Ce-V, as well as the recent works of Song et al. [112] on Ce-Ti, Zhan et al. [113] on Ce-Ti-Zr, and Zhou et al. [114] on Ce-La-Y combined conversion layers for the protection of aluminum and magnesium alloys open up an entirely new direction for the development of combined conversion coatings.

Especially in the context of aluminum-based alloys, the CeCC formation was considered as an approach for the elaboration of "stainless" aluminum by Mansfeld et al. [63] decades ago. However, the current interest in corrosion protection through CeCC coating primer deposition persists. Numerous research works on this topic attempt to be summarized in several extended reviews. In these reviews, the authors focus on the beneficial properties of CeO_2-based anticorrosion layers [115–117], as environmental chromate substitutions [118–120]. Other authors highlight the technological aspects of CeCC layer depositions [121], considering these types of layers as efficient sealants for anodic oxide layers [122, 123].

In this context, experimental activities follow the main streams, aiming to optimize CeCC deposition conditions [124–126], explore applications of Ce-oxide sealants for anodic oxide layers [104, 127–131], and utilize cerium compounds as corrosion inhibitors for the active protection of epoxy [132] and sol-gel-derived [133–139] protective coatings.

Regardless of the method of CeCC deposition, the sequence of related operations includes: (i) preliminary treatment; (ii) layer deposition procedure, and (iii) final CeCC sealing. In this context, the treatment of already deposited CeCC layers with hot phosphate solutions appears to be an attractive CeCC sealing procedure, as proposed by Heller et al. [140, 141].

Anodization and plasma electrochemical oxidation (PEO) are considered the most suitable approaches for preliminary surface treatment. Some authors define anodization as a process of electrochemical oxidation of metals and/or alloys under conditions of anodic polarization [142]. This method enables the formation of highly textured surfaces, composed of arrays of regularly ordered tubular pores with equal size dimensions, as illustrated in **Figure 4** [142].

The topology of the anodic aluminum oxide (AAO) layers depends on various factors during layer formation, including: (i) nature and concentration of the contact electrolyte [143]; (ii) applied current density or formation voltage [128, 144]; (iii) composition of the to-be-anodized aluminum alloy [145–149] and its structure [150]; (iv) temperature [151]; (v) duration [55, 152]; (vi) presence of additives [153], including Ce(IV) species [154], and of course, (vii) preliminary surface treatment [155–157]. Additionally, it should be mentioned that anodization at rather high voltages alters the process mechanism, converting it to plasma electrolytic oxidation (PEO). The above-listed parameters impact the PEO process, and the resulting surface modifications are investigated in detail by Egorkin et al. [158], Pezzato et al. [159], Cerchier et al. [160], and Kostelac et al. [161]. In this context, Arunnellaiappan et al.

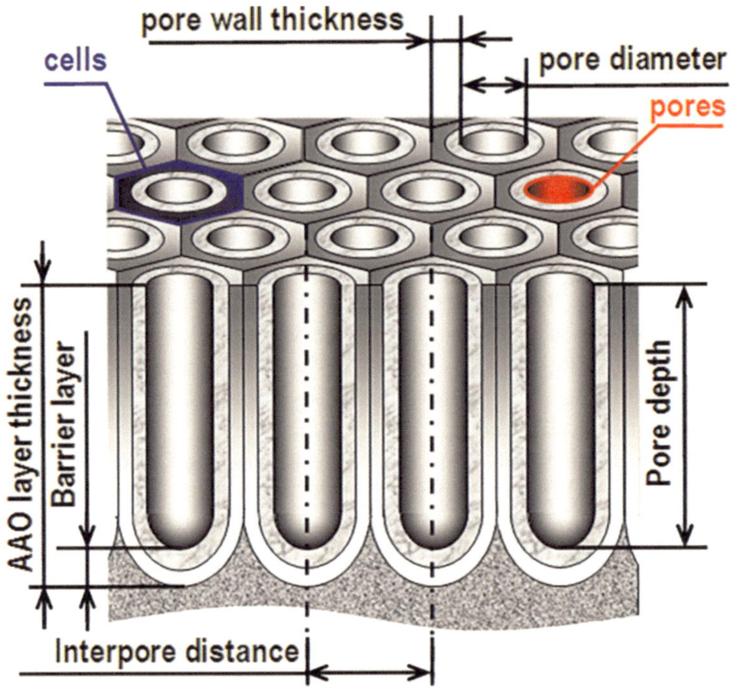

Figure 4.
Schematic illustration of a typical anodic aluminum oxide layer [142].

[162] propose the formation of combined AAO/CeCC layers by applying PEO, combined with simultaneous electrophoresis of CeO_2 nanoparticles. Furthermore, the synergism between the CeCC layer and AAO sublayer, obtained under conventional conditions, was established by Andreeva et al. [163], Kozhukharov et al. [164], and confirmed by L. Selegård et al. [165] and Ramirez et al. [166].

Finally, the term "*conversion coating*" encompasses any metal oxide layer deposited on a metal surface through chemical or electrochemical processes that partially or completely replace the native surface metal oxide, according to Arenas and Damborenea [167]. Consequently, the formation of sufficiently thick anodic aluminum oxide (AAO) layers enables their partial substitution by Ce-oxides. This substitution is possible due to the definitively acidic properties of the Ce-precursor solutions [168, 169] and the previously mentioned amphoteric character of the aluminum oxide layers, including AAO.

Although CeO_2 is a rather stable metallic oxide, commonly used as a high-performance ceramic material, it is soluble through electrochemical polarization in strongly acidic media under aeration conditions. According to the encyclopedia of Sukhotina [170], it can undergo dissolution through the following reaction:

$$CeO_2 + 4H^+ + O_2 + e^- \rightarrow Ce^{3+} + 2H_2O \quad E^0 = 1.4\ V_{SHE} \qquad (14)$$

The solubility of CeO_2 in acidic media (i.e., 0.01 M $NaClO_4$) has been thoroughly investigated by Plakhova et al. [171], comparing theoretical thermodynamic calculations with practical experiments. Additionally, Joshi et al. [172] conducted an extended experimental study on Ce-leaching from CeCC deposited on AA7075-T6 aircraft alloy in a 0.1 M NaCl model corrosive medium. In both cases, the authors claim that CeO_2 is less soluble than Ce(III) oxides and hydroxides, which constitute the CeCC layers. However, CeO_2 does possess, albeit insignificant, solubility that could be enhanced in strongly acidic media. Furthermore, the solubility of CeO_2 could be additionally enhanced when this oxide is in the form of nanoparticles [171, 172], particularly as in the CeCC layers obtained by Selegård et al. [165], composed of nanoglobules. Therefore, the post-treatment procedure is necessary for improvement of the CeCC performance in corrosive media.

In this context, phosphatation appears to be among the most attractive approaches for the final sealing of CeCC layers due to the remarkable phosphate anion adsorption aptitude of CeO_2 [173]. One of the pioneering works in this field is authored by Heller et al. [140]. The authors conducted a comparative research on the effect of $NH_4H_2PO_4$, Na_3PO_4, $K_4P_2O_7$, $Na_2H_2P_2O_7$, and $Na_5P_3O_{10}$ as phosphate sources. Furthermore, the temperature and duration of the post-treatment have been the subject of research [140, 141]. The authors immersed CeCC-coated AA2024-T3 plates in a 2.5 wt.% $NH_4H_2PO_4$ solution for different times (10, 30, 120, and 600 s) and temperatures (55, 70, and 85°C), respectively. It was concluded that the best coating is obtained for at least 10 minutes at a minimum of 70°C and pH = 4.5. Additionally, at room temperature, no changes have been observed. The authors explained the beneficial effect of the post-treatment with the formation of supplemental hydrated $CePO_4$, which fills the cracks of the CeCC coating, making it much denser. Furthermore, similar methods are proposed for the corrosion protection of zinc [174] and low carbon steel [175]. Recently, Tsanev et al. [176] proposed phosphatation of the CeCC layer deposited on AA1050 alloy by exposure for 15 min to 0.5 M NaH_2PO_4 with the addition of 0.1 M $Ca(NO_3)_2$ at 85°C. Considering the amphoteric properties of the AAO layers and the CeO_2 solubility in definitely acidic media, Portolesi et al. [177] propose two buffering solution compositions enabling the final sealing of the AAO/CeCC layers at nearly neutral pH, namely: Na_2HPO_4/KH_2PO_4 and Na_2B4O_7 / HCl. In this sense, other systems, like $Na_2B_4O_7/H_3BO_3$, are of potential interest for further neutral solutions for final thermal treatment.

Direct Ce-phosphate deposition on AA2024-T3 alloy has also been proposed [178, 179]. Recently, Wang et al. [180] introduced chemically bonded phosphate ceramic coatings modified with organic-inorganic hybrid nano alumina. Other authors suggest hydrothermal sealing of AAO layers after CeCC deposition [165] or incorporation of copper oxide [181].

Steels, as critical metallic materials in civil engineering, are under scrutiny for corrosion protection through Ce-compounds via inhibition [182–184], CeCC deposition [185–188], and the utilization of Ce-compounds as active ingredients in protective coatings [183].

Similar to aluminum alloys, the properties of the CeCC layer on steel substrates are contingent upon preliminary treatment and phosphatation. Therefore, steel substrates should undergo preliminary galvanization, as iron is susceptible to severe uniform corrosion attack in CeCC deposition solutions [189]. This susceptibility is due to the inherent acidic nature of CeCC deposition solutions and the oxidant properties of the Ce(IV) species, discussed below (**Figure 5**).

In this context, Zn-based coatings for steels can also be applied through electrochemical methods. The quality of the resulting galvanic layers depends on both the electrolyte composition [190–192], the applied galvanizing regime [191], and the post-treatment applied. This procedure can be executed through anodization [192–196], similar to the case of aluminum discussed earlier, or by the deposition of Ce-modified hydrotalcites, as proposed by Pham et al. [197, 198]. In summary, among the most promising approaches for corrosion protection could be achieved by subsequent performance of: (i) galvanization; (ii) galvanic layer anodization; (iii) CeCC layer deposition; (iv) phosphate sealing; and (v) hydrotalcite finishing, combined with (vi) advanced polymer deposition.

All electrochemical layer formation processes should be conducted after at least a brief analysis of the system to be elaborated, considering the interactions between the metal substrate and the metallic ions in the deposition electrolyte. For this reason, the standard electrode potentials, defined against the *Standard Hydrogen Electrode* (SHE), of the redox couples of ceric to cerous ions and some of the metals recently subjected to CeCC, are summarized in **Table 1** [170].

Moreover, theoretically, the Ce(IV) and Ce(III) compounds from the CeCC deposition electrolytes could also be reduced to metallic cerium:

$$Ce^{3+} + 3e- \rightarrow Ce^0 \ E^0 = -2.483 \ V_{SHE} \tag{15}$$

$$Ce^{4+} + 4e- \rightarrow Ce^0 \ E^0 = 1.68 \ V_{SHE} \tag{16}$$

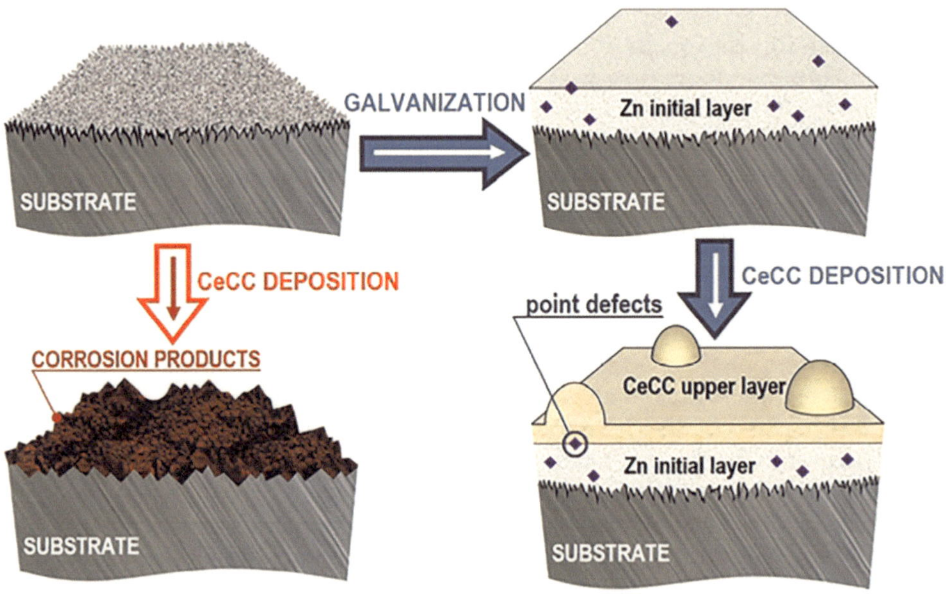

Figure 5.
Illustration of CeCC deposition on steel after preliminary Zn –galvanization according to [189].

	Oxidized/reduced couple	Standard electrode potential
Strong oxidant	Ce^{4+} + e- → Ce^{3+}	E^0 = 1.61 V_{SHE}
↕	Cu^{2+} + $2e^-$ → Cu^0 (metallic)	E^0 = 0.3472 V_{SHE}
	Fe^{2+} + $2e^-$ → Fe^0 (metallic)	E^0 = −0.4402 V_{SHE}
	Zn^{2+} + $2e^-$ → Zn^0 (metallic)	E^0 = −0.7688 V_{SHE}
	Al^{3+} + $3e^-$ → Al^0 (metallic)	E^0 = −1.662 V_{SHE}
Strong reducer	Mg^{2+} + 2e- → Mg^0 (metallic)	E^0 = −2.363 V_{SHE}

Table 1.
Standard electrode potentials of Ce^{4+}/Ce^{3+} and some metals.

However, no data are found for electrochemical or chemical metallic cerium layer deposition, probably because the conditions necessary for the formation of metallic Ce layers are extremely severe for the metals to be coated. The potentials shown in **Table 1** should be recalculated to be adopted for real conditions. For instance, the potentials of the saturated calomel (SCE) and the silver/silver chloride (Ag/AgCl 3 M KCl) reference electrodes are as follows: $E^0_{SCE} = 0.268\ V_{SHE}$ and $E^0_{Ag/AgCl} = 0.222\ V_{SHE}$.

Additionally, the real standard potential values should be recalculated using the Nernst equation and considering the Pourbaix *E*/pH diagrams, as proposed elsewhere [50]. Particularly, the CeCC formation solutions are definitely acidic, commonly with $pH \approx 2.5$ [108, 109].

Finally, the remarkable oxidant properties of the Ce^{4+} ions find application in various fields, including analytical, environmental, and nanomedicine applications [199–201]. For example, enzymes, being biocatalysts with protein-based composition and various specific activities, can be substituted with CeO_2 nanoparticles, as proposed by Xiao et al. [202]. CeO_2 nanoparticles have potential applications in oxidative nanozymes, environmental [203], medical [204], and bactericide [205] applications.

Advanced corrosion-protective coatings should be composed of various layers, each performing its own function. Coating primers, like the aforementioned CeCC, support adhesion between metallic surfaces and intermediate layers. These intermediate layers, composed of organic or hybrid polymers, are susceptible to ultraviolet (UV)-assisted chemical bond disruption or oxidation. Hence, they should be protected by outer UV-absorptive coating finishes. Both the intermediate layers and the coating finishes should possess distinguishable hydrophobicity.

Efficient advanced coating systems should respond to a variety of requirements to provide passive and active corrosion protection. The former pertains to the barrier properties of the coating layer between the protected metal and the corrosive medium. The latter relates to the residual barrier effects after the disruption of coating integrity, as summarized in a recent review [206]. Recent research activities are focused on acquiring superhydrophobicity and self-healing ability [207–212]. The illustration of recent and upcoming trends in the design of advanced corrosion-protective layers for aluminum alloys is provided below (**Figure 6**).

In this context, the sol-gel method emerges as the most suitable approach for the development of intermediate and finishing layers in advanced corrosion-protective

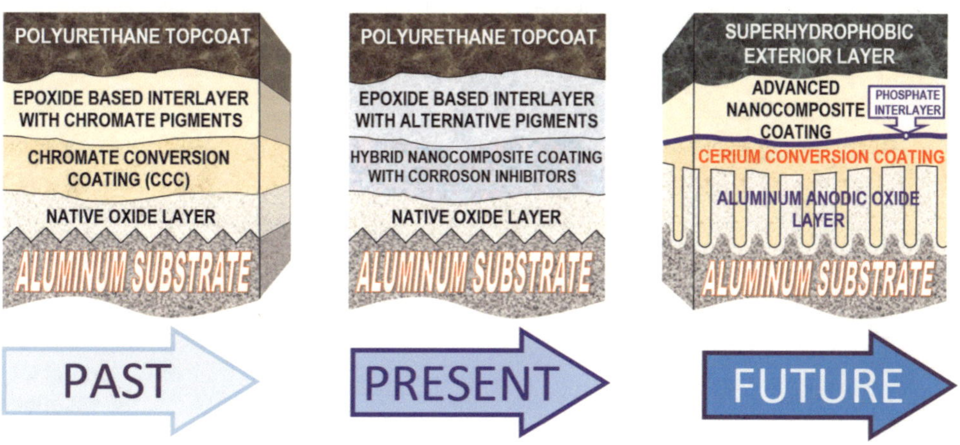

Figure 6.
Recent trends regarding the elaboration of advanced corrosion-protective coatings for aluminum alloys.

coating systems. The obtained gels represent an intermediate stage in the coating formation process, followed by appropriate drying and thermal treatment to achieve dense and adherent solid layers. Generally, gels are considered micro-heterogeneous colloidal systems, composed of a liquid medium equally distributed in a polymeric matrix [213]. The application of this approach requires monitoring various factors, including: (i) the composition and nature of the precursors, (ii) the used solvents, (iii) the medium pH, (iv) the addition of modifiers such as detergents and complexing agents, (v) temperature, and (vi) the surrounding gaseous medium composition and pressure [214, 215].

The sol-gel method allows for the direct incorporation of CeO_2 nanoparticles into the initial liquid sol. This approach leads to significantly higher scratch resistance in the resulting sol/gel-derived intermediate layers. Moreover, these particles can be pre-loaded with water-soluble Ce-compounds, such as $Ce(NO_3)_3$ [80, 215, 216], $CeCl_3$ [215], and $(CH_3COO)_3Ce$ [215], to ensure their gradual release as efficient corrosion inhibitors, as illustrated below (**Figure** 7).

This concept enables the use of a variety of corrosion inhibitors and nanocontainers [217–226], allowing for the combination of diverse polymer matrix materials, solid nanoparticle phases and inhibitors.

Specifically, the sol-gel method proves to be a versatile approach for the synthesis and deposition of functional thin films. Sol-gel-derived layers can serve various functions, being effective corrosion protectors, as previously mentioned, as well as active sensor components and electrodes for alternative energy sources.

3. Cerium compounds for sensor elements

The sensors are indispensable components essential for the operation of every monitoring and automatic control system. In classical automation, automatic systems are generally classified into *Systems for Automatic Regulation* (SAR) and *Systems for Automatic Control* (SAC). The key distinction lies in the fact that SAR controls a single parameter, while SAC manages multiple parameters simultaneously. Contemporary automation trends emphasize the use of sensor arrays for precise and reliable control. This stands in contrast to earlier generations of automatic regulation systems that heavily relied on the technical functionality of a single sensor, often without the ability to verify its status.

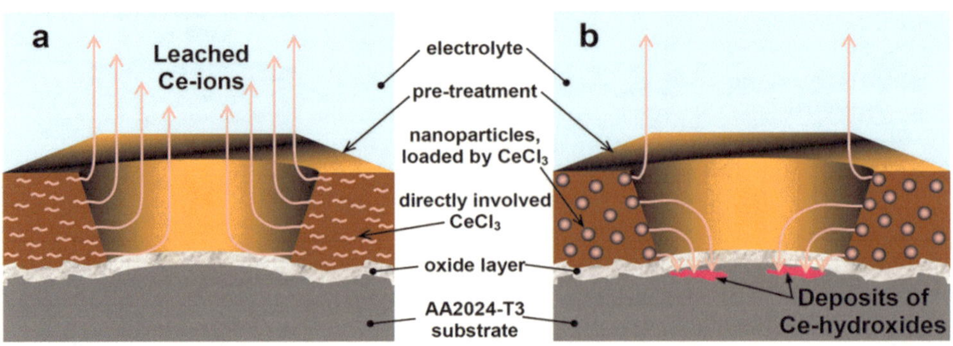

Figure 7.
3D schematic view for the role of nanocontainers for decrease of the undesirable leaching of the inhibitor; (a) – inhibitor involved directly in GMT-matrix and (b) – inhibitor-loaded nanoparticles [217].

However, new challenges arise in modern automatic systems as the use of multiple parallel sensors introduces the potential for deviations among the data provided by each sensor. In cases where significant data discrepancies emerge, it indicates a malfunction in at least one of these elements.

The term "*sensor element*" encompasses all electronic components that generate electric signals or modify their parameters as a response to their environmental conditions [227]. Sensor elements generating electric signals are termed "*primary elements*," while those relying on an external source of electric energy for their function are considered "*secondary elements*." Chemical sensors play a crucial role as interfaces between chemical industrial installations and automatic control systems. These sensors convert variables in the surrounding medium (inside tubes, chemical reactors, reservoirs, etc., where the sensor is installed) into electric output signal parameters. Examples of environmental variables and resulting output signal parameters are illustrated in **Figure 8**.

The sensors, depending on their intended use, play a crucial role in detecting and monitoring various environmental parameters, such as temperature (°t), pressure (P), concentration of a chemical compound [C], acidity or alkalinity of liquids (pH), relative humidity of gases (RH%), light intensity (candela) or wavelength (meter), etc.

The construction of these sensors determines the electric output signal parameters that include electric current (I), potential (U), ohmic resistance (R), capacitance (C), or inductance (L). The development of sensor elements involves analyzing the correlations between environmental variables and output signal parameters. In addition to sensitivity (tg(α)), these elements have inherent detection and output signal ranges, critical for detectable values of the environmental variable and the output signal.

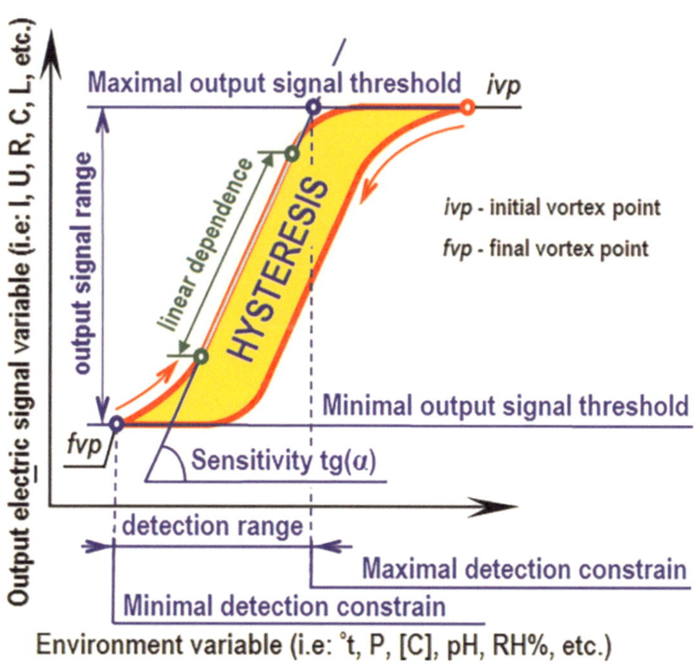

Figure 8.
Schematic illustration of sensor element characteristics [227].

Outside these ranges, the sensor does not alter the output signal characteristics and may even be damaged.

Beyond the parameters mentioned, sensors in development undergo testing for durability during storage under various conditions and reliability after extended usage. These characteristics are determined through long-term test procedures, considering data repeatability and reproducibility. Repeatability pertains to a sensor's ability to consistently perform under multiple repetitions of environmental variable alterations, while reproducibility is concerned with the similarity of performance among a relatively large number of identical sensor elements. Statistical data acquisition is employed through multiple repetitions of sensor characteristics' determination under the same conditions. This approach allows for statistical analysis, resulting in the final result (FR), which includes the respective standard deviation. Thus, the final results of multiple measurements are defined by the following formula:

$$FR = \sum_{i=1}^{i=n} \frac{AR_i}{n} \pm \sqrt{\sum_{i=1}^{i=n} \frac{(AR_i - AR_{av})^2}{n(n-1)}}; \tag{17}$$

Where: FR is the final result. AR_i is the acquired result of each measurement with i ranging from at least 2 to n. n is the total number of measurements. AR_{av} is the average value obtained by dividing the sum of all measurement results by their number.

Nonetheless, sensors undergo changes in behavior over time due to aging and/or corrosion of their components. The former (i.e., aging) are result of internal non desirable structural and compositional alterations, caused by internal factors, such as structural relaxation, residual crystallization, solid-state diffusion, etc. The latter is (i.e., corrosion) result of interactions with the surrounding medium, as was already mentioned. Additional important characteristics of sensors include response time delay and recovery time. Response time delay refers to the time between the moment of notable alteration of the controlled parameter (i.e., environmental variable) and the moment of its registration. Recovery time predetermines the hysteresis area. Both these parameters should have as lower values as possible, achievable by decreasing the size dimensions of the sensor elements, for instance, by using nanosized materials like nanocrystalline CeO_2, as proposed by Knecht et al. [228], and by employing precise equipment for sample assembly.

Chemical sensors also should exhibit selectivity regarding the substance (i.e., chemical element or compound) whose concentration should be controlled.

Nanocrystalline CeO_2-based materials hold great potential for use in sensor elements due to their recently described semiconductor properties arising from the composition of CeO_2 lattice [229–233], which has a highly defective and nonstoichiometric structure [234–236]. Utilizing the sol-gel method, humidity sensor elements based on Ce-doped titanium dioxide (TiO_2) layers were developed [237], even though obtaining Ce-Ti-O proved challenging due to the large atomic radius of Ce preventing its involvement in the TiO_2 crystal lattice. Nevertheless, the coexistence of both oxides could result in conjunction with interesting optical and electric properties. This approach was further enhanced, leading to the development of sol-gel-derived CeO_2-TiO_2-SiO_2 humidity sensors with remarkable properties [238, 239]. The introduction of CeO_2 nanoparticles into a polymer matrix is also employed for the development of advanced biosensors [240, 241].

Another approach to enhance the sensing ability of CeO_2-based materials involves modifying their lattices by doping with other elements [25, 240–242]. The zinc oxide

(ZnO) modification by Ce is particularly attractive, as the treatment of ZnO/TiO_2 sol-gel-derived film at 400°C results in the formation of the simonkolleite phase, which possesses exceptional hygroscopicity [243]. The beneficial effect of ZnO-TiO_2-SiO_2 thin films is confirmed by other authors [244–247], with the procedures proposed allowing for the involvement of Ce-compounds aiming at further improvement of the resulting film sensitivity.

4. Cerium compounds for alternative energy sources

The aforementioned specific semiconductor properties of CeO_2-based materials and the possibilities for their doping make them suitable for use in alternative energy sources, such as *solid oxide fuel cells* (SOFCs). SOFCs are solid-state multilayer ceramic electrochemical devices, composed of at least two porous electrodes separated by a dense ion-conductive electrolyte. The SOFCs enable simultaneous thermal and electric energy generation (**Figure 9a**) and efficient electrocatalytic oxidation of *volatile organic compounds* (VOCs), as electrocatalytic oxidation precursors or environmental contaminants (**Figure 9b**).

The characteristics of the output signals generated by SOFCs depend on the gas flows and concentrations. Consequently, SOFCs could function as gas sensors in the transport and industrial sectors. In transportation, SOFC sensors could detect incompletely combusted exhaust gases. In industry, they could be used for various gaseous flows of oxidants (O_2, O_3, etc.) or reductants (H_2 vapors of volatile organic compounds, such as the light fraction combustibles in the petrol refineries, etc.), as well as for detecting toxic gases (especially carbon monoxide) [248], applicable in metallurgical plants and thermal electric power stations.

In this regard, cerium oxide (CeO_2) and doped ceria-based materials are well suited for SOFC applications due to their previously mentioned electrical conductivity, oxygen adsorption ability, and extended mechanical properties [249].

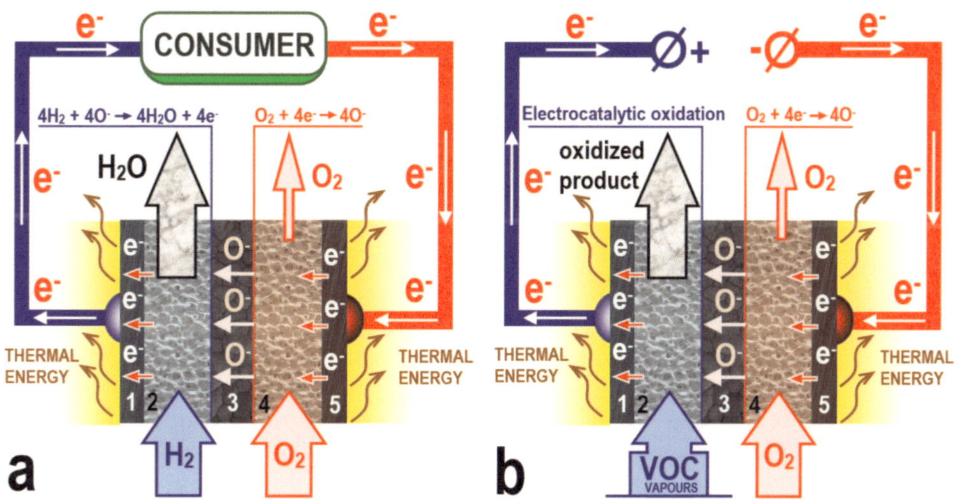

Figure 9.
Schematic illustration of SOFC, working as energy source (a) and electrocatalyst (b) 1, 5-contact elements (platinum (Pt) or stainless steel); 2-porous anode; 3-dense solid electrolyte (usually yttria-stabilized zirconia); 4-porous cathode.

Additionally, their conductivity and other properties can be easily modified by varying their structure and porosity [250, 251]. Further improvement can be achieved when CeO_2 is doped with other elements, such as La [252], Gd [253–255], In [256], Y, Zr [257], Fe, Cu [258, 259], etc.

The hydrogen required for feeding SOFCs can be provided by photoelectrochemical water-splitting electrolysers [260]. These devices compensate for the low efficiency of classical electrolysers by utilizing sunlight energy captured by photosensitive electrodes. In this context, attention is focused on CeO_2 materials doped with Cu [261], Fe [262, 263], Ni [263–265], Se [264, 265], Co [266, 267], etc.

In this context, CeO_2 phosphatation appears to be a rather attractive approach, not only for efficient corrosion protection, as discussed in the previous paragraph, but also for the development of photoactive electrodes for water-splitting [268–270].

Finally, as CeO_2-based materials belong to the class of metal oxides, their synthesis is possible by drop spreading of solutions and/or colloids containing Ce-salts across high-temperature workspaces or on heated surfaces. This approach is well known as *spray pyrolysis*. In the former case, this method enables the *spray pyrolysis synthesis* (SPS) of nanosized powders [271]. In the latter case, *spray pyrolysis deposition* (SPD) allows the formation of thin CeO_2 layers [272], facilitating the development of advanced devices such as sensor elements [273]. The CeO_2-based nanomaterials have recently encountered other applications for energy storage, like supercapacitors [274–278], environmental applications [279], Zn-air batteries [280, 281], etc.

Moreover, the unique properties of CeO_2-based materials are under investigation for potential applications as superconductors [282], magnetic materials [283], and proton ceramic membrane reactors [284]. Additionally, spray freezing methods have been developed as an alternative to the spray pyrolysis approach [285]. Other alternative methods for CeO_2-based nanomaterials synthesis are the hydrothermal method, proposed by *Hunpratub* et al. [277], and gel combustion method, proposed recently by various authors [286–289].

5. Conclusions

Recently, interest in the application of Ce-based materials has dramatically increased due to their unique properties and relatively reasonable economic accessibility. Cerium compounds can be employed in various forms for efficient corrosion protection of various metallic substrates, as modern protective coatings comprise multiple layers with distinct functions.

For advanced coating primers on Al-based alloys, a process involving Al substrate anodization, CeCC deposition, and phosphate sealing can be employed.

Similarly, coating primers for steels can be developed through initial galvanizing, followed by CeCC deposition and final phosphatation.

Cerium compounds can also be utilized for intermediate and finishing coating layers, providing both passive and active corrosion protection. In the former case, CeO_2 can serve as a reinforcing phase to enhance the mechanical properties of the coating. In the latter case, various water-soluble Ce-compounds (mainly salts) can function as corrosion inhibitors, acting even after coating integrity is compromised.

The specific semiconductive properties exhibited by CeO_2 make this oxide particularly intriguing for the development of advanced sensor elements.

This oxide is extensively studied as part of alternative energy sources, such as solid oxide fuel cells (SOFCs) and photoelectrochemical water splitters. SOFCs generate electricity and thermal energy while consuming hydrogen, and photoelectrochemical water splitters produce hydrogen essential for SOFC operation. Additionally, Ce-based SOFCs can be applied for electrocatalysis of reactions for various chemical and environmental purposes. Besides, the nanosized ceria can also find applications in biochemical fields, serving as a nanozyme with potential medical applications. Furthermore, ceria has demonstrated utility in superconductors and magnetic materials.

Several promising technological methods, including the sol-gel method, spray pyrolysis, spray freezing, hydrothermal and gel combustion methods, are considered as prospective approaches for ceria-based nanosized materials synthesis.

Acknowledgements

This work is developed as part of contract №: BG-RRP-2.004-0002-C01, project name: BiOrgaMCT, Procedure BG-RRP-2.004 "Establishing of a network of research higher education institutions in Bulgaria," funded by Bulgarian National Recovery and Resilience Plan. Also, the authors are grateful for the funding of this research to the Bulgarian National Science Fund contract No. КП-06-ПН79/1 "Protective and functional layers on aluminum and its alloys."

Author details

Stephan Kozhukharov*, Christian Girginov, Vanya Lilova and Plamen Petkov
University of Chemical Technology and Metallurgy, Sofia, Bulgaria

*Address all correspondence to: s.kozhukharov@uctm.edu

References

[1] Falconnet PJ. The rare earth industry: A world of rapid change. Journal of Alloys and Compounds. 1993;**192**: 114-117. DOI: 10.1016/0925-8388(93) 90203-Y

[2] Hughes AE, Mol JMC, Cole IS. The cost and availability of rare earth-based corrosion inhibitors. In: Forsyth M, Hinton B, editors. Rare Earth-based Corrosion Inhibitors. Woodhead Publishing; 2014. p. 319. DOI: 10.1533/ 9780857093585.291

[3] Muecke GK, Möller P. The not-so-rare earths. Scientific American. 1988; **258**:72-78. Available from: https://www.jstor.org/stable/24988948

[4] Haley TJ. Pharmacology and toxicology of the rare earth elements. Journal of Pharmaceutical Sciences. 1965;**54**:633-670. DOI: 10.1002/ jps.2600540502

[5] Sweet DV, Fang J, Shiang D. Registry of toxic effects of chemical substances. Registry of Toxic Effects of Chemical Substances (RTECS). United States: Computer Tape. Data files; 1986. p. 103

[6] Strmcnik D, Papa LP, Genorio B, Stamenkovic VR, Markovic NM. Design principles for hydrogen evolution reaction catalyst materials. Nano Energy. 2016;**29**:29-36. DOI: 10.1016/j. nanoen.2016.04.017

[7] Shao Y, Markovic NM. Prelude: The renaissance of electrocatalysis. Nano Energy. 2016;**29**:1-3. DOI: 10.1016/j. nanoen.2016.09.025

[8] Haeussler A, Abanades S. Two-step thermochemical cycles using fibrous ceria pellets for H_2 production and CO_2 reduction in packed-bed solar reactors. Sustainable Materials and Technologies. 2021;**29**:e00328. DOI: 10.1016/j. susmat.2021.e00328

[9] Haeussler A, Abanades S, Julbe A, Jouannaux J, Drobek M, Ayral A, et al. Remarkable performance of microstructured ceria foams for thermochemical splitting of H_2O and CO_2 in a novel high-temperature solar reactor. Chemical Engineering Research and Design. 2020;**156**: 311-323. DOI: 10.1016/j.cherd.2020. 02.008

[10] Haeussler A, Abanades S, Julbe A, Jouannaux J, Cartoixa B. Solar thermochemical fuel production from H_2O and CO_2 splitting via two-step redox cycling of reticulated porous ceria structures integrated in a monolithic cavity-type reactor. Energy. 2020;**201**: 117649. DOI: 10.1016/j.energy.2020. 117649

[11] Oliveira FAC, Barreiros MA, Haeussler A, Caetano APF, Mouquinho AI, Silva PMO, et al. High-performance cork-templated ceria for solar thermochemical hydrogen production via two-step water-splitting cycles. Sustainable Energy & Fuels. 2020;**4**:3077-3089. DOI: 10.1039/ D0SE00318B

[12] Ranganathan RV, Jony B, Fondriest SM, Liu Z, Wang R, Uddi M. Plasma-catalysis chemical looping CH_4 reforming with water splitting using ceria supported Ni based La-perovskite nano-catalyst. Journal of CO_2 Utilization. 2019;**32**:11-20. DOI: 10.1016/j.jcou. 2019.03.014

[13] Orfila M, Sanz D, Linares M, Molina R, Sanz R, Marugán J, et al. H_2 production by thermochemical water splitting with reticulated porous

structures of ceria-based mixed oxide materials. International Journal of Hydrogen Energy. 2021;**46**(33): 17458-17471. DOI: 10.1016/j.ijhydene. 2020.04.222

[14] Sahu AK, Zhao XS, Upadhyayula S. Ceria-based photocatalysts in water-splitting for hydrogen production and carbon dioxide reduction. Catalysis Reviews. DOI: 10.1080/01614940. 2023.2166227

[15] Wang B, Li X, Dai Y, Wang C-H. Thermodynamic analysis of an epitrochoidal rotary reactor for solar hydrogen production via a water-splitting thermochemical cycle using nonstoichiometric ceria. Energy Conversion and Management. 2022;**268**: 115968. DOI: 10.1016/j.enconman. 2022.115968

[16] Wang P, Wei RK, Vafai K. A dual-scale transport model of the porous ceria based on solar thermochemical cycle water splitting hydrogen production. Energy Conversion and Management. 2022;**272**:116363. DOI: 10.1016/j. enconman.2022.116363

[17] Abanades S, Haeussler A, Julbe A. Synthesis and thermochemical redox cycling of porous ceria microspheres for renewable fuels production from solar-aided water-splitting and CO_2 utilization. Applied Physics Letters. 2021;**119**:023902. DOI: 10.1063/5.0055282

[18] Onigbajumo A, Swarnkar P, Will G, Sundararajan T, Taghipour A, Couperthwaite S, et al. Techno-economic evaluation of solar-driven ceria thermochemical water-splitting for hydrogen production in a fluidized bed reactor. Journal of Cleaner Production. 2022;**371**:133303. DOI: 10.1016/j. jclepro.2022.133303

[19] Woo S, Lee J, Lee DS, Kim JK, Lim B. Electrospun carbon nanofibers with embedded Co-Ceria nanoparticles for efficient hydrogen evolution and overall water splitting. Materials. 2020;**13**(4): 856. DOI: 10.3390/ma13040856

[20] Shaghaghi Z, Jafari S, Mohammad-Rezaei R. The heterostructure of ceria and hybrid transition metal oxides with high electrocatalytic performance for water splitting and enzyme-free glucose detection. Journal of Electroanalytical Chemistry. 2022;**915**:116369. DOI: 10.1016/j.jelechem.2022.116369

[21] Salinas D, Pecchi G, Fierro JLG. K_2O supported on sol-gel CeO_2-Al_2O_3 and La_2O_3-Al_2O_3 catalysts for the transesterification reaction of canola oil. Journal of Molecular Catalysis A: Chemical. 2016;**A-423**:503-510. DOI: 10.1016/j.molcata.2016.08.002

[22] Ratthanaphra D, Suwanmanee U. Uncertainty analysis of environmental sustainability of biodiesel production using Thai domestic rare earth oxide solid catalysts. Sustainable Production and Consumption. 2019;**18**:237-249. DOI: 10.1016/j.spc.2019.01.001

[23] Takte MA, Ingle NN, Dole BN, Tsai M-L, Hianik T, Shirsat MD. A stable and highly-sensitive flexible gas sensor based on Ceria (CeO_2) nano-cube decorated rGO nanosheets for selective detection of NO_2 at room temperature. Synthetic Metals. 2023;**297**:117411. DOI: 10.1016/j. synthmet.2023.117411

[24] Baier D, Priamushko T, Weinberger C, Kleitz F, Tiemann M. Selective discrimination between CO and H_2 with copper–ceria-resistive gas sensors. ACS Sensors. 2023;**8**(4): 1616-1623. DOI: 10.1021/acssensors. 2c02739

[25] Akçay N, Algün G, Öztel HO. Fabrication of ultra-sensitive humidity sensors based on Ce-doped ZnO

nanostructure with superfast response and recovery time. Journal of Materials Science: Materials in Electronics. 2023; **34**:1565. DOI: 10.1007/s10854-023-10973-y

[26] Song Y, Meng X, Bi M, Gao W. Conductometric H_2 sensors at room temperature based on palladium nanoparticles on ceria. Sens. Actuators. 2023;**B-375**(15):132957. DOI: 10.1016/j.snb.2022.132957

[27] Paul R, Ghosh R. Ceria quantum dot-decorated carbon nanotubes as a room-temperature acetone sensor. ACS Applied Nano Materials. 2023;**6**(12): 10223-10235. DOI: 10.1021/acsanm.3c01029

[28] Abdulwahab KO, Khan MM, Jennings JR. Doped ceria nanomaterials: Preparation, properties, and uses. ACS Omega. 2023;**8**(34):30802-30823. DOI: 10.1021/acsomega.3c01199

[29] Feng N, Liu Y, Dai X, Wang Y, Guo Q, Li Q. Advanced applications of cerium oxide based nanozymes in cancer. RSC Advances. 2022;**12**: 1486-1493. DOI: 10.1039/D1RA05407D

[30] Mohamed RMK, Mohamed SH, Asran AM, Alsohaimi IH, Hassan HMA, Ibrahim H. Carbon microspheres uniformly decorated with ceria nanoparticles as an ultrasensitive platform for electrochemical sensing of antihypertensive drug lacidipine in patient plasma and pharmaceutical formulation. Microchemical Journal. 2023;**187**:108422. DOI: 10.1016/j.microc.2023.108422

[31] Naik YK, Sharma V. Fabrication of paper based cerium oxide (CeO_2) square disc sensor for detection of standard antioxidants. The Pharma Innovation Journal. 2023;**12**(6):3555-3558. Available from: https://www.thepharmajournal.com/archives/2023/vol12issue6/PartAO/12-6-232-897.pdf

[32] Liaqat F, Haq I, Saira F, Qaisa S. Development of glucose sensor based on cobalt and nickel doped ceria nanostructures. Materials Science and Engineering. 2023;**B-289**:116231. DOI: 10.1016/j.mseb.2022.116231

[33] Ma B, Han J, Zhang K, Jiang Q, Sui Z, Zhang Z, et al. Targeted killing of tumor cells based on isoelectric point suitable nanoceria-rod with high oxygen vacancies. Journal of Materials Chemistry. 2022;**B-10**(9):1410-1417. DOI: 10.1039/D1TB02787E

[34] Li H, Xi P, Pan S, Qi Z, Fu C, Yu Z, et al. The Advances of Ceria Nanoparticles for Biomedical Applications in Orthopaedics. International Journal of Nanomedicine. 2020;**15**:7199-7214. DOI: 10.2147/IJN.S270229

[35] Mehmood R, Mofarah SS, Rawal A, Tomasetig F, Wang X, Yang J-L, et al. Green synthesis of zwitterion-functionalized nano-octahedral ceria for enhanced intracellular delivery and cancer therapy. ACS Sustainable Chemistry & Engineering. 2019;**7**(10): 9189-9201. DOI: 10.1021/acssuschemeng.8b06726

[36] Khan M, Mashwani Z-R, Ikram M, Raja NI, Mohamed AH, Ren G, et al. Efficacy of green cerium oxide nanoparticles for potential therapeutic applications: circumstantial insight on mechanistic aspects. Nanomaterials. 2022;**12**(12):2117. DOI: 10.3390/nano12122117

[37] Panda SR, Singh RK, Priyadarshini B, Rath PP, Parhi PK, Sahoo T, et al. Nanoceria: A rare-earth nanoparticle as a promising anti-cancer therapeutic agent

in colon cancer. Materials Science in Semiconductor Processing. 2019;**104**: 104669. DOI: 10.1016/j.mssp.2019.104669

[38] Datta A, Mishra S, Manna K, Saha KD, Mukherjee S, Roy S. Pro-oxidant therapeutic activities of cerium oxide nanoparticles in colorectal carcinoma cells. ACS Omega. 2020;**5**(17): 9714-9723. DOI: 10.1021/acsomega.9b04006

[39] Sridharan M, Kamaraj P, Vennilaraj C, Arockiaselvi J, Pushpamalini T, Vivekanand PA, et al. Synthesis, characterization and evaluation of biosynthesized Cerium oxide nanoparticle for its anticancer activity on breast cancer cell (MCF 7). Materials Today. 2021;**36**:914-919. DOI: 10.1016/j.matpr.2020.07.031

[40] Johnson KK, Kopecky C, Koshy P, Liu Y, Devadason M, Holst J, et al. Theranostic activity of ceria-based nanoparticles toward parental and metastatic melanoma: 2D vs 3D Models. ACS Biomaterials Science & Engineering. 2023;**9**(2):1053-1065. DOI: 10.1021/acsbiomaterials.2c01258

[41] Rozhin P, Melchionna M, Fornasiero P, Marchesan S. Nanostructured ceria: Biomolecular templates and bioapplications. Nanomaterials. 2021;**11**(9):2259. DOI: 10.3390/nano11092259

[42] Georgantzia E, Gkantou M, Kamaris GS. Aluminium alloys as structural material: A review of research. Engineering Structures. 2021;**227**:111372. DOI: 10.1016/j.engstruct.2020.111372

[43] da Silva AAM, Arruti E, Janeiro G, Aldanondo E, Alvarez P, Echeverria A. Material flow and mechanical behaviour of dissimilar AA2024-T3 and AA7075-T6 aluminium alloys friction stir welds. Materials & Design. 2011;**32**:2021-2027. DOI: 10.1016/j.matdes.2010.11.059

[44] Yasakau KA, Zheludkevich ML, Lamaka SV, Ferreira MGS. Mechanism of corrosion inhibition of AA2024 by rare-earth compounds. The Journal of Physical Chemistry. 2006;**B110**:5515-5528. DOI: 10.1021/jp0560664

[45] Roberge PR. Handbook of Corrosion Engineering. Chapter 5. McGraw-Hill; 2000. pp. 331-369

[46] Kozhukharov S, Milanes M, Girginov C, Machkova M. Comparative evaluation of cerium oxide primers electrodeposited on AA2024-T3 and D16 AM aircraft alloys. Materials and Corrosion. 2016;**67**(7):710-720. DOI: 10.1002/maco.201508635

[47] Mardaras E, Imaz N, González-Martínez R, García-Lecina E, Aguado E. Influence of the casting process on corrosion behaviour of 2024-T6 aluminium alloy in different conditions of pH and NaCl concentration. Corrosion Engineering, Science and Technology. 2023;**58**(3):214-222. DOI: 10.1080/1478422X.2022.2162661

[48] Olgiati M, Denissen PJ, Garcia SJ. When all intermetallics dealloy in AA2024-T3: Quantifying early stage intermetallic corrosion kinetics under immersion. Corrosion Science. 2021;**192**: 109836. DOI: 10.1016/j.corsci.2021.109836

[49] Kosari A, Tichelaar F, Visser P, Zandbergen H, Terryn H, Mol JMC. Dealloying-driven local corrosion by intermetallic constituent particles and dispersoids in aerospace aluminium alloys. Corrosion Science. 2020;**177**: 108947. DOI: 10.1016/j.corsci.2020.108947

[50] Kozhukharov SV, Girginov CA. Classical and modern methods for corrosion impact rate determination for aluminium and strengthened aircraft alloys. Fundamentals and practical applications. In: Gergely A, editor. Phenomena and Theories in Corrosion Science. Methods of Prevention. New York: NOVA Sci. Publ.; 2019. pp. 3-150

[51] Zhang X, Zhou X, Hashimoto T, Liu B. Localized corrosion in AA2024-T351 aluminium alloy: Transition from intergranular corrosion to crystallographic pitting. Materials Characterization. 2017;**130**: 230-236. DOI: 10.1016/j.matchar.2017.06.022

[52] Katona RM, Karasz EK, Schaller RF. A review of the governing factors in pit-to-crack transitions of metallic structures. Corrosion. 2023;**79**(1):72-96. DOI: 10.5006/4179

[53] Alexopoulos ND, Velonaki Z, Stergiou CI, Kourkoulis SK. The effect of artificial ageing heat treatments on the corrosion-induced hydrogen embrittlement of 2024 (Al–Cu) aluminium alloy. Corrosion Science; 2016;**2016**(102):413-424. DOI: 10.1016/j.corsci.2015.10.034

[54] Gangloff RP. Environmental Cracking-Corrosion Fatigue. In: Baboian R, editor. Corrosion tests and standards manual. 2nd ed. West Conshohocken, PA (USA): ASTM International; 2005. pp. 302-321

[55] Girginov C, Kozhukharov S, Milanes M, Machkova M. Impact of the anodizing duration on the surface morphology and performance of A2024-T3 in a model corrosive medium. Materials Chemistry and Physics. 2017; **198**:137-144. DOI: 10.1016/j.matchemphys.2017.05.049

[56] Iwai M, Kikuchi T. Chemical stability of porous anodic aluminum oxide in both acidic and alkaline solutions. Thin Solid Films. 2023;**771**: 139784. DOI: 10.1016/j.tsf.2023.139784

[57] Miyamoto K, Iwai M, Kikuchi T. Rapid electrochemical separation of anodic porous alumina films from aluminum surfaces using a highly safe sodium chloride–ethylene glycol solution. Electrochimica Acta. 2022;**427**: 140865. DOI: 10.1016/j.electacta.2022.140865

[58] Hinton BRW, Arnott DR, Ryan NE. The inhibition of aluminium corrosion by cerium cations. Metals Forum. 1984;**7**: 211-217

[59] Hinton BRW, Arnott DR, Ryan NE. Cerium conversion coating for the corrosion protection of aluminium. Metals Forum. 1986;**9**:162-173

[60] Davenport AJ, Isaacs HS, Kendig MW. X-Ray Absorption Study of Cerium in the Passive Film on Aluminum. Journal of the Electrochemical Society. 1989;**136**: 1837-1838

[61] Davenport AJ, Isaacs HS, Kendig MW. XANES investigation of the role of cerium compounds as corrosion inhibitors for aluminum. Corrosion Science. 1991;**32**(5–6):653-663

[62] Mansfeld F, Wang Y, Shih H. The Ce-Mo process for the development of a stainless aluminum. Electrochimica Acta. 1992;**37**(12):2277-2282

[63] Mansfeld F, Wang Y. Development of "stainless" aluminum alloys by surface modification. Materials Science and Engineering. 1995;**A-198**:51-61

[64] Scholes FH, Soste C, Huges AE, Hardin SG, Curtis PR. The role of

hydrogen peroxide in the deposition of cerium-based conversion coatings. Applied Surface Science. 2006;**253**: 1770-1780

[65] Krainova DA, Zharkinova ST, Saetova NS, Raskovalov AA, Kuzmin AV, Eremin VA, et al. Influence of cerium oxide on properties of glass–ceramic sealants for solid oxide fuel cells. Russian Journal of Applied Chemistry. 2017;**90**:1278-1284. DOI: 10.1134/S1070427217080146

[66] de Oliveira MAC, D'Epifanio A, Ohnuki H, Meche B. Platinum group metal-free catalysts for oxygen reduction reaction: Applications in microbial fuel cells. Catalysts. 2020;**10**(475):1-22. DOI: 10.3390/catal10050475

[67] Pernas JE, Kozhukharov S, Salve AA, Matter E, Machkova M. Influence of the oxidation state of Ce-ions on the inhibition of AA2024 alloy corrosion in a model corrosive medium. Journal of the University of Chemical Technology and Metallurgy. 2012;**47**(3):311-318

[68] Bethencourt M, Botana FJ, Calvino JJ, Marcos M, Rodriguez-Chacon MA. Lanthanide compounds as environmentally-friendly corrosion inhibitors of aluminium alloys: A review. Corrosion Science. 1998;**40**:1803-1819. DOI: 10.1016/S0010-938X(98)00077-8

[69] Jegdić B, Bobić B, Linić S. Corrosion behaviour of AA2024 aluminium alloy in different tempers in NaCl solution and with the $CeCl_3$ corrosion inhibitor. Materials and Corrosion. 2020; **71**(3):352-3641. DOI: 10.1002/maco.201911219

[70] Kurtela M, Šimunović V, Stojanović I, Alar V. Effect of the cerium (III) chloride heptahydrate on the corrosion inhibition of aluminum alloy. Materials and Corrosion. 2020;**71**(1): 125-147. DOI: 10.1016/j.porgcoat.2019.105428

[71] Pejić JN, Jegdić BV, Radojković BM, Marunkić DD, Marinković AD, Bajat JB. Cysteine and cerium as green corrosion inhibitors for AA7049: Mixture vs. complexation. Corrosion Science. 2022; **209**:110815. DOI: 10.1016/j.corsci.2022.110815

[72] Volarič B, Milošev I. Rare earth chloride and nitrate salts as individual and mixed inhibitors for aluminium alloy 7075-T6 in chloride solution. Corrosion Engineering, Science and Technology. 2017;**52**(3):201-2113. DOI: 10.1080/1478422X.2016.1245957

[73] Jegdić B, Bobić B, Radojković B, Kovačina J, Marunkić D. Synergistic effect of $CeCl_3$ and benzotriazole on corrosion resistance of naturally aged and artificially aged AA2024 aluminium alloy. Transactions of Nonferrous Metals Society of China. 2020;**30**(6):1478-1490. DOI: 10.1016/S1003-6326(20)65312-2

[74] Radojković B, Marunkić D, Milošević M, Grubišić S, Pejić J, Jegdić B, et al. Analysis of the Ce(III) chloride, Ce (III) nitrate, and Ce(III) propionate as corrosion inhibitor of the aluminum alloys in NaCl solution. Materials and Corrosion. 2022;**73**(10):1604-1614. DOI: 10.1002/maco.202213177

[75] Zheng T, Wang L, Liu J. Corrosion inhibition of levofloxacin and $Ce(NO_3)_3$ for AA2024-T4 in 3.5% NaCl. Corrosion Engineering, Science and Technology. 2020;**55**(1):75-82. DOI: 10.1080/1478422X.2019.1681152

[76] Tianyu Z, Lu W, Jinyan L. Corrosion inhibition of levofloxacin and $Ce(NO_3)_3$ for AA2024-T4 in 3.5% NaCl. Corrosion Engineering, Science and Technology. 2020;**55**(1):75-82. DOI: 10.1080/1478422X.2019.1681152

[77] Lopez-Sesenes R, Gonzalez-Rodriguez JG, Vera-Dimas JG, Guardian-Tapia R, Cisneros-Villalobos L. Organic and inorganic compounds as corrosion inhibitors to reduce galvanic effect for the hybrid structure AA2024-CFPR. Journal of Electrochemical Science and Engineering. 2022;**12**(2):343-358. DOI: 10.5599/jese.1126

[78] Zhu C, Yang HX, Wang YZ, Zhang DQ, Chen Y, Gao LX. Synergistic effect between glutamic acid and rare earth cerium (III) as corrosion inhibitors on AA5052 aluminum alloy in neutral chloride medium. Ionics. 2019;**25**: 1395-1406. DOI: 10.1007/s11581-018-2605-4

[79] Snihirova D, Lamaka SV, Taheri P, Mol JMC, Montemor MF. Comparison of the synergistic effects of inhibitor mixtures tailored for enhanced corrosion protection of bare and coated AA2024-T3. Surface and Coating Technology. 2016;**303-B**:342-351. DOI: 10.1016/j.surfcoat.2015.10.075

[80] El-Hashemy MA, Hughes AE, Gengenbach T, Glenn AM, Cole IS. Combined influence of Ce(III) and iodide ions for corrosion protection of AA2024-T3 in acidic to neutral chloride-rich environments: Electrochemical and surface characterization studies. Journal of Rare Earths. 2023;**41**:309-320. DOI: 10.1016/j.jre.2022.05.014

[81] Gobara M, Baraka A, Akid R, Zorainy M. Corrosion protection mechanism of Ce^{4+}/organic inhibitor for AA2024 in 3.5% NaCl. RSC Advances. 2020;**10**(4):2227-2240. DOI: 10.1039/c9ra09552g

[82] Byrne CE, CNV Q, D'Alessandro O, Deyá C. Cerium and aluminum molybdates as inhibitors of the aluminum AA1050 corrosion process in aqueous NaCl solutions. Materials and Corrosion. 2024;**75**:73-83. DOI: 10.1002/maco.202313940

[83] Rodic P, Lekka M, Andreatta F, Milošev I, Fedrizzi L. The synergistic effect of cerium acetate and sodium sulphate on corrosion inhibition of AA2024-T3 at various temperatures. Electrochimica Acta. 2021;**370**:137664. DOI: 10.1016/j.electacta.2020.137664

[84] Rodic P, Milosev I, Lekka M, Francesco A, Fedrizzi L. Study of the synergistic effect of cerium acetate and sodium sulphate on the corrosion inhibition of AA2024-T3. Electrochimica Acta. 2019;**308**(10):337-349. DOI: 10.1016/j.electacta.2019.04.042

[85] Simović AR, Radojković BM, Jegdić BV, Milojević M, Bajat J. Cerium lactate as green corrosion inhibitor of AA2024 alloy. Materials and Corrosion. 2023;**74**(9):1371-1381. DOI: 10.1002/maco.202313882

[86] Marunkić D, Pejić J, Jegdić B, Linić S, Perišić J, Radojković B, et al. Inhibitory effect of cerium salts of lower carboxylic acids on Al-Zn-Mg-Cu alloy in NaCl solution. Journal of the Electrochemical Society. 2021;**168**(8): 081501. DOI: 10.1149/1945-7111/ac1895

[87] Umoren SA, Solomon MM. Synergistic corrosion inhibition effect of metal cations and mixtures of organic compounds: A Review. Journal of Environmental Chemical Engineering. 2017;**5**:246-273. DOI: 10.1016/j.jece.2016.12.001

[88] Panchal J, Shah D, Patel R, Shah S, Prajapati M, Shah M. Comprehensive review and critical data analysis on corrosion and emphasizing on green eco-friendly corrosion inhibitors for oil and gas industries. Journal of Bio- and Tribo-Corrosion. 2021;7:107. DOI: 10.1007/s40735-021-00540-5

[89] Zehra T, Kaseem M. Lanthanides: The key to durable and sustainable corrosion protection. ACS Sustainable Chemistry & Engineering. 2023;**11**(18): 6776-6800. DOI: 10.1021/ acssuschemeng.3c00763

[90] Khan MAA, Irfan OM, Djavanroodi F, Asad M. Development of sustainable inhibitors for corrosion control. Sustainability (Switzerland). 2022;**14**(15):9502. DOI: 10.3390/ su14159502

[91] Ma IAW, Ammar S, Kumar SSA, Ramesh K, Ramesh S. A concise review on corrosion inhibitors: types, mechanisms and electrochemical evaluation studies. Journal of Coating Technology and Research. 2022;**19**(1): 241-268. DOI: 10.1007/s11998-021-00547-0

[92] Zhang Q, Shen M, Liu X, Fu H, Yang W, Zhao J, et al. Construction and utilization of rare earth complexes as efficient corrosion inhibitor in epoxy coating. Progress in Organic Coating. 2024;**186**:108024. DOI: 10.1016/j. porgcoat.2023.108024

[93] Sanni O, Ren J, Jen T-C. Agro-industrial wastes as corrosion inhibitor for 2024-T3 aluminum alloy in hydrochloric acid medium. Results in Engineering. 2022;**16**: 100676. DOI: 10.1016/j.rineng. 2022.100676

[94] Conde A, Arenas MA, de Frutos A, de Damborenea J. Effective corrosion protection of 8090 alloy by cerium conversion coatings. Electrochimica Acta. 2008;**53**:7760-7768

[95] Harchegani RK, Riahi AR. Effect of cerium chloride on the self-corrosion and discharge activity of aluminum anode in alkaline aluminum-air batteries. Journal of the Electrochemical Society. 2022;**169**(31):030542. DOI: 10.1149/ 1945-7111/ac2476

[96] Kozhukharov S, Girginov C. Recent trends of the use of rare earth elements for efficient environmentally compliant corrosion protection of aluminum and its alloys. In: Petkov P, Achour ME, Popov C, editors. NATO Science for Peace and Security Series B: Physics and Biophysics. Dordrecht, Netherlands: Springer; 2020. pp. 437-445. ISBN: 978-94-024-2044-9

[97] Rodríguez DS, Kozhukharov S, Machkova M, Kozhukharov V. Influence of the deposition conditions on the properties of D16 AM clad alloy, dip-coated in Ce-containing baths. Bulgarian Chemical Communications. 2013;**45-A**: 24-32

[98] Kozhukharov S, Ayuso JAP, Rodríguez DS, Acuña OF, Machkova M, Kozhukharov V. Optimization of the basic parameters of cathodic deposition of Ce-conversion coatings on D16 AM clad alloy. Journal of Chemical Technology and Metallurgy. 2013;**48**(3): 296-307

[99] Ayuso JAP, Kozhukharov S, Machkova M, Kozhukharov V. Electrodeposition of cerium conversion coatings for corrosion protection of D16 AM clad alloy. Bulgarian Chemical Communications. 2013;**45-A**:33-40

[100] Gil TP, Kozhukharov S, Girginov C. Deposition of cerium conversion coatings on AA2024-T3 aircraft alloy at fixed potentials. In: Annual Proceedings of "Angel Kanchev". Vol. 53. Bulgaria: University of Ruse; 2014. pp. 9-13. Available from: https://conf.uni-ruse.bg/ bg/docs/cp14/10.1/10.1-1.pdf

[101] Živković LJS, Popić JP, Jegdić BV, Dohčević-Mitrović Z, Bajat JB, Mišković-

Stanković VB. Corrosion study of ceria coatings on AA6060 aluminum alloy obtained by cathodic electrodeposition: Effect of deposition potential. Surface and Coating Technology. 2014;**240**: 327-335. DOI: 10.1016/j.surfcoat. 2013.12.048

[102] Yu X, Li G. XPS study of cerium conversion coating on the anodized 2024 aluminum alloy. Journal of Alloys and Compounds. 2004;**364**:193-198. DOI: 10.1016/S0925-8388(03)00502-4

[103] Zhang S, Wang C, Zhao S, Niu A, Ma Y, Liu B. Enhanced long-term corrosion protection of 2A14 aluminum alloy: Hybrid effect of micro-arc oxidation coating and cerium based conversion treatment. Surface and Coating Technology. 2023;**464**:129579. DOI: 10.1016/j.surfcoat.2023.129579

[104] Covelo A, Rodil S, Novoa XR, Hernandez M. Development and characterization of sealed anodizing as a corrosion protection for AA2024-T3 in saline media. Materials Today Communications. 2022;**31**:103468. DOI: 10.1016/j.mtcomm.2022.103468

[105] Palomino LEM, Aoki IV, de Melo HG. Microstructural and electrochemical characterization of Ce conversion layers formed on Al alloy 2024-T3 covered with Cu-rich smut. Electrochimica Acta. 2006;**51**:5943-5953. DOI: 10.1016/j.electacta.2006.03.036

[106] Hughes AE, Scholes FH, Glenn AM, Lau D, Muster TH, Hardin SG. Factors influencing the deposition of Ce-based conversion coatings, part I: The role of Al3+ ions. Surface and Coating Technology. 2009;**203**:2927-2936. DOI: 10.1016/j.surfcoat.2009.03.022

[107] Kozhukharov S, Acuña OF, Machkova MS, Kozhukharov V. Influence of buffering on the spontaneous deposition of cerium conversion coatings for corrosion protection of AA2024-T3 aluminum alloy. Journal of Applied Electrochemistry. 2014;**44**:1093-1105

[108] Kozhukharov S, Girginov C, Boshkova N, Tzanev A. Impact of the final thermal sealing of combined Zinc/ Cerium oxide protective coating primers formed on low carbon steel. Journal of Electrochemical Science and Engineering. 2022;**12**(4):685-701. DOI: 10.5599/jese.1297

[109] Guergova D, Stoyanova E, Stoychev D, Avramova I, Atanasova G, Stefanov P. Corrosion stability of stainless steel, modified electrochemically with Ce_2O_3 – CeO_2 films, in 3.5% NaCl media. Bulgarian Chemical Communications. 2011;**43**(1): 150-157

[110] Aramaki K. Self-healing mechanism of a protective film prepared on a $Ce(NO_3)_3$-pretreated zinc electrode by modification with $Zn(NO_3)_2$ and Na_3PO_4. Corrosion Science. 2003;**45**: 1085-1101. DOI: 10.1016/S0010-938X (02)00153-1

[111] Jiang X, Guo R, Jiang S. Evaluation of self-healing ability of Ce–V conversion coating on AZ31 magnesium alloy. Journal of Magnesium and Alloys. 2016;**4**(3):230-241. DOI: 10.1016/j. jma.2016.06.003

[112] Song D, Ma M, Zhao L. Study of sealing of anodized aluminium in mixed titanium-cerium salt solutions. International Journal of Electrochemical Science. 2022;**17**:221224. doi. DOI: 10.20964/2022.12.23

[113] Zhan W, Qian X, Gui B, Liu L, Liu X, Li Z, et al. Preparation and corrosion resistance of titanium–zirconium–cerium based conversion

coating on 6061 aluminum alloy. Materials and Corrosion. 2020; **71**(3):419-4291. DOI: 10.1002/maco.201911193

[114] Zhou Y, Zhang P, Su L, Fan B, Yan F. One-step preparation and characterization of a Ce-La-Y ternary rare earth conversion coating on magnesium alloy AZ91D. Materials Letters. 2021;**304**:130640. DOI: 10.1016/j.matlet.2021.130640

[115] Rajendran P, Muthuraj A, Rajagounder NE. Review on CeO_2-based corrosion coatings. Transactions of the Indian Ceramic Society. 2022;**81**(4): 158-174. DOI: 10.1080/0371750X.2022.2149623

[116] Castano CE, O'Keefe MJ, Fahrenholtz WG. Cerium-based oxide coatings. Current Opinion in Solid State & Materials Science. 2015; **19**:69-76. DOI: 10.1016/j.cossms.2014.11.005

[117] Presuel-Moreno F, Jakab MA, Tailleart N, Goldman M, Scully JR. Corrosion-resistant metallic coatings. Materials Today. 2008;**11**:15-23. DOI: 10.1016/S1369-7021(08)70203-7

[118] Prakash BS, Balaraju JN. Chromate (Cr^{6+})-free surface treatments for active corrosion protection of aluminum alloys: A review. Journal of Coating Technology and Research. 2024;**21**:105-135. DOI: 10.1007/s11998-023-00831-1

[119] Peltier F, Thierry D. Review of Cr-free coatings for the corrosion protection of aluminum aerospace alloys. Coatings. 2022;**12**(4):518. DOI: 10.3390/coatings12040518

[120] Becker M. Chromate-free chemical conversion coatings for aluminum alloys. Corrosion Reviews. 2019;**37**(4):321-342. DOI: 10.1515/corrrev-2019-0032

[121] Harvey TG. Cerium-based conversion coatings on aluminium alloys: A process review. Corrosion Engineering, Science and Technology. 2013;**48**:248-269. DOI: 10.1179/1743278213Y.0000000089

[122] Ofoegbu SU, Feernandes FAO, Pereira AB. The sealing step in aluminum anodizing: A focus on sustainable strategies for enhancing both energy efficiency and corrosion resistance. Coatings. 2020;**10**(3):226. DOI: 10.3390/coatings10030226

[123] Lkhagvaa T, Rehman ZU, Choi D. Post-anodization methods for improved anticorrosion properties: a review. Journal of Coating Technology and Research. 2021;**18**:1-17. DOI: 10.1007/s11998-020-00367-8

[124] Hinder SJ, Grilli R, Rustame M, Santos WIA, Baker MA, Costa I. A surface analytical investigation of cerium-based conversion coatings deposited onto an AA2024-T3 aluminium alloy cladding layer. Surface and Interface Analysis. 2014;**46**(10–11): 735-739. DOI: 10.1002/sia.5440

[125] Milošev I, Volarič B. Conversion coatings based on rare earth nitrates and chlorides for corrosion protection of aluminum alloy 7075-T6. Corrosion. 2017;**73**(7):822-843. DOI: 10.5006/2353

[126] Klumpp RE, Araujo JVS, Antunes RA, Gonçalves de Viveiros BV, Magnani M, Costa I. A cerium-based nanocoating for corrosion protection of the AA1230 as clad material for the AA2024-T3 alloy. Materials Research. 2022;**25**(Suppl. 1):e20210616. DOI: 10.1590/1980-5373-MR-2021-0616

[127] Fedel M, Franch J, Rossi S. Effect of thickness and sealing treatments on the corrosion protection properties of anodic oxide coatings on AA5005. Surface and

Coating Technology. 2021;**40825**:126761. DOI: 10.1016/j.surfcoat.2020.126761

[128] Nerímio NR, Casademont C, Arurault L. Influence of the anodization electrical mode on the final properties of electrocolored and sealed anodic films prepared on 1050 aluminum alloy. Materials Chemistry and Physics. 2022; **288**:126369. DOI: 10.1016/j.matchemphys.2022.126369

[129] Wang Q, Li H. Study on anodic oxidation of 2099 aluminum lithium alloy and sealing treatment in environmental friendly solutions. International Journal of Electrochemical Science. 2023;**18**(7):100186. DOI: 10.1016/j.ijoes.2023.100186

[130] Queiroz FM, Bugarin AFS, Ayusso VH, Terada M, Costa I. Investigation on the effect of a chromium-free sealing treatment for the corrosion resistance of AA2198-T851 after tartaric sulphuric anodizing (TSA). Corrosion and Materials Degradation. 2023;**4**(2):331-344. DOI: 10.3390/cmd4020017

[131] Prada Ramirez OM, Queiroz FM, Tunes AM, Antunes RA, Rodrigues CL, Lanzutti A, et al. Tartaric-sulphuric acid anodized clad AA2024-T3 post-treated in Ce-containing solutions at different temperatures: Corrosion behaviour and Ce ions distribution. Applied Surface Science. 2020;**53430**:147634. DOI: 10.1016/j.apsusc.2020.147634

[132] Hu T, Shi H, Fan S, Liu F, Han E-H. Cerium tartrate as a pigment in epoxy coatings for corrosion protection of AA 2024-T3. Progress in Organic Coating. 2017;**105**:123-131. DOI: 10.1016/j.porgcoat.2017.01.002

[133] Liu W, An C, Hao J, Li W. Cerium doped trimethoxy silane-aluminium isopropoxide coatings for enhanced corrosion protection of 1061 aluminum alloy in aqueous sodium chloride solution. International Journal of Electrochemical Science. 2021;**16**(3):210352. DOI: 10.20964/2021.03.18

[134] Bahrami M, Borhani GH, Bakhshi SR, Ghasemi A. Preparation and evaluation of corrosion behavior of GPTMS–TEOS hybrid coatings containing Zr and Ce on aluminum alloy 6061–T6. Journal of Sol-Gel Science and Technology. 2015;**76**(3):552-561. DOI: 10.1007/s10971-015-3805-2

[135] Bahrami M, Borhani GH, Bakhshi SR, Ghasemi A. Optimization of effective processing parameters of hybrid anti-corrosion Si/Zr sol-gel coatings doped with cerium salt for aluminum alloy 6061. Journal of Sol-Gel Science and Technology. 2017;**81**(3):921-933. DOI: 10.1007/s10971-016-4255-1

[136] Yu M, Liu Y, Liu J, Li S, Xue B, Zhang Y, et al. Effects of cerium salts on corrosion behaviors of Si-Zr hybrid sol-gel coatings. Chinese Journal of Aeronautics. 2015;**28**(2):600-608. DOI: 10.1016/j.cja.2015.01.011

[137] Thai TT, Trinh AT, Olivier MG. Hybrid sol–gel coatings doped with cerium nanocontainers for active corrosion protection of AA2024. Progress in Organic Coating. 2020;**138**:105428. DOI: 10.1016/j.porgcoat.2019.105428

[138] Yasakau KA, Kuznetsova A, Kallip S, Starykevich M, Tedim J, Ferreira MGS, et al. A novel bilayer system comprising LDH conversion layer and sol-gel coating for active corrosion protection of AA2024. Corrosion Science. 2018;**143**:299-313. DOI: 10.1016/j.corsci.2018.08.039

[139] Manasa S, Jyothirmayi A, Siva T, Sathiyanarayanan S, Gobi KV, Subasri R.

Effect of inhibitor loading into nanocontainer additives of self-healing corrosion protection coatings on aluminum alloy A356.0. Journal of Alloys and Compounds. 2017;**726**:969-977. DOI: 10.1016/j.jallcom.2017.08.037

[140] Heller DK, Fahrenholtz WG, O'Keefe MJ. Effect of phosphate source on post-treatment of cerium-based conversion coatings on Al 2024-T3. Journal of the Electrochemical Society. 2009;**156**(11):C400. DOI: 10.1149/1.3224005

[141] Heller DK, Fahrenholtz WG, O'Keefe MJ. The effect of post-treatment time and temperature on cerium-based conversion coatings on Al 2024-T3. Corrosion Science. 2010;**52**:360-368. DOI: 10.1016/j.corsci.2009.09.023

[142] Girginov CA, Kozhukharov SV. In: Hermancen A, editor. Chapter 1. Theoretical Bases and Mechanisms of the Electrochemical Alumina Films Formation. New York: Nova Science Publishers; 2020. pp. 3-108

[143] Kozhukharov S, Girginov C, Avramova I, Machkova M. Anodic galvanostatic polarization of AA2024-T3 aircraft alloy in conventional mineral acids. Materials Chemistry and Physics. 2016;**180**:301-313. DOI: 10.1016/j.matchemphys.2016.06.011

[144] Tamburrano A, De Vivo B, Höijer M, Arurault L, Tucci V, Fontorbes S, et al. Effect of electric field polarization and temperature on the effective permittivity and conductivity of porous anodic aluminium oxide membranes. Microelectronic Engineering. 2011;**88**:3338-3346. DOI: 10.1016/j.mee.2011.08.007

[145] Giffard K, Arurault L, Blanc C. Dynamic measurements and wettability phenomena in mesoporous anodic films prepared on 1050 and 2024T3 aluminium alloys. Microporous and Mesoporous Materials. 2016;**235**:32-41. DOI: 10.1016/j.micromeso.2016.07.042

[146] Zaraska L, Wierzbicka E, Kurowska-Tabor E, Sulka GD. Synthesis of nanoporous anodic alumina by anodic oxidation of low purity aluminum substrates. In: Losic D, Santos A, editors. Nanoporous Alumina. Springer Series in Materials Science. Vol. 219. Cham: Springer; 2015. pp. 61-106. DOI: 10.1007/978-3-319-20334-8_3

[147] Martínez-Viademonte MP, Abrahami ST, Hack T, Burchardt M, Terryn H. A review on anodizing of aerospace aluminum alloys for corrosion protection. Coatings. 2020;**10**:1-30. DOI: 10.3390/coatings10111106

[148] Giffard K, Arurault L, Blanc C, Di Caprio D. Accurate evaluations of both porosity and tortuosity of anodic films grown on rolled AA 1050 and on rolled or machined AA 2024 T3. Surface and Interface Analysis. 2019;**51**:1184-1193. DOI: 10.1002/sia.6606

[149] Theohari S, Kontogeorgou C. Study of electrochemical behavior of commercial AA5052 during anodizing in phosphoric acid solution in relation to Mg species content in films. Surface Engineering and Applied Electrochemistry. 2020;**56**:71-82. DOI: 10.3103/S1068375520010159

[150] de Sousa Araujo JV, Milagre M, Costa I. A historical, statistical and electrochemical approach on the effect of microstructure in the anodizing of Al alloys: A review. Critical Reviews in Solid State and Materials Sciences. DOI: 10.1080/10408436.2023.2230250

[151] Chernyakova K, Tzaneva B, Vrublevsky I, Videkov V. Effect of aluminum anode temperature on growth

rate and structure of nanoporous anodic alumina. Journal of the Electrochemical Society. 2020;**167**:103506. DOI: 10.1149/1945-7111/ab9d65

[152] Raffin F, Echouard J, Volovitch P. Influence of the anodizing time on the microstructure and immersion stability of tartaric-sulfuric acid anodized aluminum alloys. Metals. 2023;**13**(5):993. DOI: 10.3390/met13050993

[153] Kaur H, Sharma L, Kushwaha M. Effect of anionic surfactant on the growth of anodic nanoporous aluminum oxide. ECS Journal of Solid State Science and Technology. 2016;**5**:M154-M157. DOI: 10.1149/2.0151612jss

[154] Del Olmo R, Mohedano M, Visser P, Rodriguez A, Matykina E, Arrabal R. Effect of cerium (IV) on thin sulfuric acid anodizing of 2024-T3 alloy. Journal of Materials Research and Technology. 2021;**15**:3240-3254. DOI: 10.1016/j.jmrt.2021.09.117

[155] Matter EA, Kozhukharov SV, Machkova MS. Effect of preliminary treatment on the superficial morphology and the corrosion behaviour of AA2024 aluminum alloy. Bulgarian Chemical Communications. 2011;**43**(1):23-30

[156] Matter EA, Kozhukharov S, Machkova M, Kozhukharov V. Reproducibility of the corrosion parameters for AA2024-T3 aluminium alloy in chloride solution after different preliminary treatment procedures. Journal of Chemical Technology and Metallurgy. 2015;**50**(1):52-64

[157] Chankitmunkong S, Eskin D, Limmaneevichitr C, Kengkla N, Diewwanit O. Characterization of the anodic film and corrosion resistance of an A535 aluminum alloy after intermetallics removal by different etching time. Metals. 2022;**12**(7):1140. DOI: 10.3390/met12071140

[158] Egorkin VS, Medvedev IM, Sinebryukhov SL, Vyaliy IE, Gnedenkov AS, Nadaraia KV, et al. Atmospheric and marine corrosion of PEO and composite coatings obtained on Al-Cu-Mg aluminum alloy. Materials. 2020;**13**:2739. DOI: 10.3390/ma13122739

[159] Pezzato L, Gennari C, Franceschi M, Brunelli K. Influence of silicon morphology on direct current plasma electrolytic oxidation process in AlSi10Mg alloy produced with laser powder bed fusion. Scientific Reports. 2022;**12**:14329. DOI: 10.1038/s41598-022-18176-x

[160] Cerchier P, Pezzato L, Gennari C, Moschin E, Moro I, Dabalà M. PEO coating containing copper: A promising anticorrosive and antifouling coating for seawater application of AA 7075. Surface and Coating Technology. 2020;**393**:125774. DOI: 10.1016/j.surfcoat.2020.125774

[161] Kostelac L, Pezzato L, Settimi AG, Franceschi M, Gennari C, Katya B, et al. Investigation of hydroxyapatite (HAP) containing coating on grade 2 titanium alloy prepared by plasma electrolytic oxidation (PEO) at low voltage. Surfaces and Interfaces. 2022;**30**:101888. DOI: 10.1016/j.surfin.2022.101888

[162] Arunnellaiappan T, Ashfaq M, Krishna LR, Rameshbabu N. Fabrication of corrosion-resistant Al_2O_3–CeO_2 composite coating on AA7075 via plasma electrolytic oxidation coupled with electrophoretic deposition. Ceramics International. 2016;**42**:5897-5905. DOI: 10.1016/j.ceramint.2015.12.136

[163] Andreeva RA, Stoyanova EA, Tsanev AS, Stoychev DS. Effect of cerium ions on the protective ability of

Al_2O_3 films formed anodically during their sealing in aqueous solutions. Bulgarian Chemical Communications. 2017;**49**:5-14

[164] Kozhukharov S, Girginov C. Enhancement of the cerium oxide primer layers deposited on AA2024-T3 aircraft alloy by preliminary anodization. Journal of Electrochemical Science and Engineering. 2018;**8**(2):113-127. DOI: 10.5599/jese.478

[165] Selegård L, Poot T, Eriksson P, Palisaitis J, Persson POÅ, Hu Z, et al. In-situ growth of cerium nanoparticles for chrome-free, corrosion resistant anodic coatings. Surface and Coating Technology. 2021;**410**:126958. DOI: 10.1016/j.surfcoat.2021.126958

[166] Ramirez OMP, Tunes MA, Mennucci MM, Starykevich M, Neves C, Ferreira MGS, et al. Ce post-treatment for increased corrosion resistance of AA2024-T3 anodized in tartaric-sulfuric acid. Corrosion Science. 2022;**204**(1): 110371. DOI: 10.1016/j.corsci.2022.110371

[167] Arenas M, de Damborenea J. Generation of conversion coatings using rare earths on galvanised steel. Revista de Metalurgia. 2005;**41**(Extra):433-437. DOI: 10.3989/revmetalm.2005.v41.iExtra.1070

[168] Atanasova P, Kozhukharov S, Milanes M. Evaluation of the buffering effect possessed by diluted diammonium hexanitrocerate solutions. Annual Proceedings of the University of Rousse (Bulgaria). 2015;**54**:42-46. Available from: http://conf.uni-ruse.bg/bg/docs/cp15/10.1/10.1-8.pdf

[169] Yu F, Li WF, Chen XP, Zhang GG. The mechanism of solution acidification during chemical conversion process on aluminium alloys. Materials Research Innovations. 2015;**19**:S2-28-S2-34. DOI: 10.1179/1432891715Z.0000000001310

[170] Sukhotina AM. Manual of Electrochemistry, "Chemistry" Gov. St. Petersburg; 1981. p. 154

[171] Plakhova TV, Romanchuk AY, Yakunin SN, Dumas T, Demir S, Wang S, et al. Solubility of Nanocrystalline Cerium Dioxide: Experimental Data and Thermodynamic Modeling. Journal of Physical Chemistry C. 2016;**120**(39):22615-22626. DOI: 10.1021/acs.jpcc.6b05650

[172] Joshi S, Kulp EA, Fahrenholtz WG, O'Keefe MJ. Dissolution of cerium from cerium-based conversion coatings on Al 7075-T6 in 0.1 M NaCl solutions. Corrosion Science. 2012;**60**:290-295. DOI: 10.1016/j.corsci.2012.03.023

[173] Wu B, IMC L. Surface functional group engineering of CeO_2 particles for enhanced phosphate adsorption. Environmental Science and Technology. 2020;**54**(7):4601-4608. DOI: 10.1021/acs.est.9b06812

[174] Ivanova DI. Phosphating of zinc surfaces in zinc-calcium solutions. Bulgarian Chemical Communications. 2011;**43**:54-59

[175] Tumbaleva Y, Ivanova D, Fachikov L. Effect of the P_2O_5:NO_3 – ratio on the zinc phosphate coating formation. Journal of the University of Chemical Technology and Metallurgy. 2011;**46**:357-362

[176] Tsanev A, Andreeva R, Stoychev D. Influence of the chemical composition of ceria conversion coatings, sealed in solution of NaH_2PO_4 and $Ca(NO_3)_2$, on the corrosion behavior of aluminum. Materials. 2023;**16**(19): 6499. DOI: 10.3390/ma16196499

[177] Portolesi S, Kozhukharov S, Haralanova T, Girginov C. Acidity adjustment of the solutions for coating deposition and further sealing by Lourier buffers. Proceedings of the University of Ruse (Bulgaria). 2022;**61**(10.1):68-73. Art, No: SAT-LCR-P-2-CT(R)-11. Available from: https://conf.uni-ruse.bg/bg/docs/cp22/10.1/10.1-13.pdf

[178] Heller DK, Fahrenholtz WG, Fair GE, O'Keefe MJ. Directly deposited cerium phosphate coatings for the corrosion protection of Al 2024-T3. ECS Transactions. 2010;**28**(24):203-215. DOI: 10.1149/1.3496432

[179] Markley TA, Forsyth M, Hughes AE. Corrosion protection of AA2024-T3 using rare earth diphenyl phosphates. Electrochimica Acta. 2007; **52**(12):4024-4031. DOI: 10.1016/j.electacta.2006.11.028

[180] Wang J, Wu M, Miao X, Bian D, Wang Y, Zhao Y. Chemically bonded phosphate ceramic coatings with self-healing capability for corrosion resistance. Surface and Coating Technology. 2023;**473**:129987. DOI: 10.1016/j.surfcoat.2023.129987

[181] Yakovleva NM, Kokatev AN, Oskin KI, Stepanova KV, Shulga AM. Study of black protective-decorative nanocomposite anodic coatings on the surface of AMg5 aluminum alloy. Industrial Laboratory. Diagnostics of Materials. 2023;**89**(7):34-44. (In Russ.). DOI: 10.26896/1028-6861-2023-89-7-34-44

[182] Bourenane N, Hamlaoui Y, Pedraza F. Electrochemical investigation of ceria nanoparticles as a corrosion inhibitor for steel in alkaline medium. Materials and Corrosion. 2023;**74**(9): 1356-1370. DOI: 10.1002/maco.202313821

[183] Shahini MH, Mohammadloo HE, Ramezanzadeh B. Recent advances in steel surface treatment via novel/green conversion coatings for anti-corrosion applications: a review study. Journal of Coating Technology and Research. 2022; **19**:159-199. DOI: 10.1007/s11998-021-00466-0

[184] El-Hashemy MA, Easton CD, Seeber A, Cole IS, Hughes AE. Electrochemical and surface characterisation of carbon steel exposed to mixed Ce and iodide electrolytes. Metals. 2023;**13**(9):1553. DOI: 10.3390/met13091553

[185] Arrighi C, Nguyen TT, Paint Y, Savall C, Coelho LB, Creus J, et al. Study of Ce(III) as a potential corrosion inhibitor of Zn-Fe sacrificial coatings electrodeposited on steel. Corrosion Science. 2022;**200**:110249. DOI: 10.1016/j.corsci.2022.110249

[186] Khast F, Saybani M, AAS D. Effects of copper and manganese cations on cerium-based conversion coating on galvanized steel: Corrosion resistance and microstructure characterizations. Journal of Rare Earths. 2022;**40**(6): 1002-1006. DOI: 10.1016/j.jre.2021.07.015

[187] Kozhukharov S, Girginov C, Tsanev A, Boshkova N. Synergistic effect of Zn electrodeposition and cerium conversion coating on the corrosion performance of low carbon steel. Applied Surface Science. 2022;**602**: 154254. DOI: 10.1016/j.apsusc.2022.154254

[188] Bahremand F, Shahrabi T, Ramezanzadeh B, Hosseini SA. Sustainable development of an effective anti-corrosion film over the St12-steel surface against seawater attacks using Ce(III) ions/tri-sodium phosphate anions. Scientific Reports. 2023;**13**:12169. DOI: 10.1038/s41598-023-38540-9

[189] Boshkov N. Galvanic Zn–Mn alloys—electrodeposition, phase composition, corrosion behaviour and protective ability. Surface and Coating Technology. 2003;**172**: 217-226. DOI: 10.1016/S0257-8972(03) 00463-8

[190] Pereira NM, Pereira CM, Araújo JP, Silva AF. Electrodeposition of Mn and Mn-Sn alloy using choline chloride-based ionic liquids. Journal of the Electrochemical Society. 2017;**164**:D486. DOI: 10.1149/2.1421707jes

[191] de Carvalho MF, Barbano EP, Carlos IA. Influence of disodium ethylenediaminetetraacetate on zinc electrodeposition process and on the morphology, chemical composition and structure of the electrodeposits. Electrochimica Acta. 2013;**109**:798-808. DOI: 10.1016/j.electacta.2013.07.149

[192] Ohba M, Scarazzato T, Espinosa DCR, Panossian Z. Study of metal electrodeposition by means of simulated and experimental polarization curves: Zinc deposition on steel electrodes. Electrochimica Acta. 2019; **309**:86-103. DOI: 10.1016/j.electacta.2019.04.074

[193] Bianchin ACV, Kunst SR, Mueller LT, Ferreira JZ, Morisso F, Carone CLP, et al. Influence of the anodization process on Zamak 5 corrosion resistance. Materials Research. 2020;**23**(4):e20190601. DOI: 10.1590/1980-5373-MR-2019-0601

[194] Kunst SR, Bianchin ACV, Mueller LT, Santana JA, Volkmer TM, Morisso FDP, et al. Model of anodized layers formation in Zn-Al (Zamak) aiming to corrosion resistance. Journal of Materials Research and Technology. 2021;**12**:831-847. DOI: 10.1016/j.jmrt.2021.03.027

[195] Engelkemeier K, Sun A, Voswinkel D, Grydin O, Schaper M, Bremse W. Zinc anodizing: Structural diversity of anodic zinc oxide controlled by the type of electrolyte. ChemElectroChem. 2021;**8**(12): 2155-2168. DOI: 10.1002/celc.202100216

[196] da Costa CD, Kunst SR, Tamires LT, Schneider EL, Oliveira CT. Anodized Zamac sealing. Revista Matéria. 2023;**28**(3):e20230068. DOI: 10.1590/1517-7076-RMAT-2023-0068

[197] Pham TT, Nguyen TD, Nguyen AS, Gonon M, Belfiore A, Paint Y, et al. Influence of solution pH on the structure formation and protection properties of ZnAlCe hydrotalcites layers on hot-dip galvanized steel. Surface and Coating Technology. 2023;**472**:129918. DOI: 10.1016/j.surfcoat.2023.129918

[198] Pham TT, Nguyen TD, Nguyen AS, Paint Y, Gonon M, To TXH, et al. Influence of solution pH on the structure formation and protection properties of ZnAlCe hydrotalcites layers on hot-dip galvanized steel. Surface and Coatings Technology. 2023; **472**:129918

[199] Yilin Z, Li H, Wang Y, Wang Y, Huang Z, Su H, et al. CeO2 nanoparticle transformation to nanorods and nanoflowers in acids with boosted oxidative catalytic activity. ACS Applied Nano Materials. 2021;**4**(2):2098-2107. DOI: 10.1021/acsanm.0c03387

[200] Zhang J, Wang J, Liao J, Lin Y, Zheng C, Liu J. In situ fabrication of nanoceria with oxidase-like activity at neutral pH: Mechanism and boosted bio-nanozyme cascades. ACS Applied Materials & Interfaces. 2021;**13**(42): 50236-50245. DOI: 10.1021/acsami.1c14831

[201] Lewandowska H, Wójciuk K, Karczmarczyk U. Metal nanozymes: New horizons in cellular homeostasis regulation. Applied Sciences. 2021; **11**(19):9019. DOI: 10.3390/app11199019

[202] Xiao G, Li H, Zhao Y, Wei H, Li J, Su H. Nanoceria-based artificial nanozymes: Review of materials and applications. ACS Applied Nano Materials. 2022;**5**(10):14147-14170. DOI: 10.1021/acsanm.2c03009

[203] Miri P, Karbhal I, Satnami ML, Jena VK, Ghosh S. β-cyclodextrin stabilized nanoceria for hydrolytic cleavage of paraoxon in aqueous and cationic micellar media. ACS Applied Bio Materials. 2023;**6**(4):1488-1494. DOI: 10.1021/acsabm.2c01030

[204] Nele V, Tedeschi V, Campani V, Ciancio R, Angelillo A, Graziano SF, et al. Cerium-doped self-assembling nanoparticles as a novel anti-oxidant delivery system preserving mitochondrial function in cortical neurons exposed to ischemia-like conditions. Antioxidants. 2023;**12**(2): 358. DOI: 10.3390/antiox12020358

[205] Zhu W, Wang L, Li Q, Jiao L, Yu X, Gao X, et al. Will the bacteria survive in the CeO_2 nanozyme-H_2O_2 system? Molecules. 2021;**26**(12):3747. DOI: 10.3390/molecules26123747

[206] Kozhukharov S, Samichkov VI, Girginov CA, Machkova MS. Actual trends in the elaboration of advanced multifunctional coating systems for the efficient protection of lightweight aircraft alloys. Corrosion Reviews. 2017; **35**(6):383-396. DOI: 10.1515/corrrev-2017-0026

[207] Semiletov AM, Kudelina AA, Kuznetsov YI. A superhydrophobic coating and corrosion inhibitors as a combined method for the protection of aluminum alloys in chloride solutions. International Journal of Corrosion and Scale Inhibition. 2023;**12**(3):1348-1362. DOI: 10.17675/2305-6894-2023-12-3-28

[208] Semiletov AM, Kudelina AA, Kuznetsov YI. New prospects in the application of superhydrophobic coatings and corrosion inhibitors. International Journal of Corrosion and Scale Inhibition. 2022;**11**(3): 1388-1400. DOI: 10.17675/2305-6894-2022-11-3-28

[209] Zhang H, Guo Z. Recent advances in self-healing superhydrophobic coatings. NanoToday. 2023;**51**:101933. DOI: 10.1016/j.nantod.2023.101933

[210] del Olmo R, Tiringer U, Milošev I, Visser P, Arrabal R, Matykina E, et al. Hybrid sol-gel coatings applied on anodized AA2024-T3 for active corrosion protection. Progress in Organic Coating. 2021;**419**:127251. DOI: 10.1016/j.surfcoat.2021.127251

[211] Chen L, Ni X, Shen Y, Liu Z, Liu C. Experimental and simulation investigation on hydrophobicity and corrosion resistance of graphene oxide reinforced composite coating. Applied Surface Science. 2024;**648**:159072. DOI: 10.1016/j.apsusc.2023.159072

[212] Yibo O, Wei Y, Zhang R, Li R, Lin Z, Shi S, et al. Enhancing corrosion inhibition of zinc with biomimetic slippery liquid-infused porous surfaces (SLIPS): An on-site fabrication strategy. Colloids and Surfaces. 2024;**A-681**: 132779. DOI: 10.1016/j.colsurfa.2023.132779

[213] Kozhukharov S. Relationship between the conditions of preparation by the sol-gel route and the properties of the obtained products. Journal of University of Chemical Technology and Metallurgy. 2009;**44**(9):143-150

[214] Chen C, Yu M, Zhan Z, Ge Y, Sun Z, Liu J. Effect of pH on the structure and corrosion protection properties of sol-gel coatings. Corrosion Science. 2023;**212**:110995. DOI: 10.1016/j.corsci.2022.110955

[215] Rodič P, Milošev I, Frankel GS. Corrosion of Synthetic Intermetallic Compounds and AA7075-T6 in Dilute Harrison's Solution and Inhibition by Cerium(III) Salts. Journal of the Electrochemical Society. 2023; **170**(3):031503. DOI: 10.1149/1945-7111/acc0a3

[216] Nguyen TT, Arrighi C, Thai TT, Dangreau L, Gonon MF, Trinh AT, et al. Inhibitive effect of the Ce (III) chloride and nitrate on the corrosion resistance of Zn alloyed sacrificial coatings: Effect of alloying compounds of the sacrificial layer. Electrochimica Acta. 2023;**452**: 142296. DOI: 10.1016/j.electacta.2023.142296

[217] Matter E, Kozhukharov S, Machkova M, Kozhukharov V. Influence of the interactions between the corrosion inhibitor and nanocontainers over the corrosion protective capability of hybrid nanocomposite pre-treatments. Annual Proceedings of the University of Ruse (Bulgaria). 2009;**48**(9):19-23. Available from: https://conf.uni-ruse.bg/bg/docs/cp09/9/9-3.pdf

[218] Zhang HH, Zhang X, Bian H, Zhang L, Chen Y, Yang Y, et al. Benzotriazole loaded CeO_2 nano-containers towards superior anti-corrosive silane coating for protection of copper. Colloids and Surfaces. 2024;**682**: 132844. DOI: 10.1016/j.colsurfa.2023.132844

[219] Zehra S, Mobin M, Aslam R, Bhat SI. Nanocontainers: A comprehensive review on their application in the stimuli-responsive smart functional coatings. Progress in Organic Coating. 2023;**176**:107389. DOI: 10.1016/j.porgcoat.2022.107389

[220] Nóvoa XR, Pérez C. The use of smart coatings for metal corrosion control. Current Opinion in Electrochemistry. 2023;**40**:101324. DOI: 10.1016/j.coelec.2023.101324

[221] Abdolah ZM, van der Zwaag S, Garcia SJ. Self-healing corrosion-protective sol–gel coatings based on extrinsic and intrinsic healing approaches. In: Hager M, van der Zwaag S, Schubert U, editors. Self-healing Materials. Advances in Polymer Science, vol. 273. Cham: Springer; 2016. pp. 185-218. DOI: 10.1007/12_ 2015_339

[222] Figueira RB. Hybrid sol–gel coatings for corrosion mitigation: A critical review. Polymers. 2020;**12**:689. DOI: 10.3390/polym12030689

[223] Gulicovski J, Bajat J, Jokić B, Panić V, Mišković-Stanković V, Milonjić S. Protective ability and impedance response of sol–gel reversely transformed ceria conversion coating on aluminium. Journal of Solid State Electrochemistry. 2016;**20**:293-303. DOI: 10.1007/s10008-015-3040-3

[224] Sousa RPCL, Figueira RB, Callone E, Dirè S, Costa SPG, Raposo MMM. Amino-alcohol organic-inorganic hybrid sol-gel materials based on an epoxy bicyclic silane: Synthesis and characterization. Nanomaterials. 2023;**13**:2429. DOI: 10.3390/nano13172429

[225] Bijapur K, Molahalli V, Shetty A, Toghan A, De Padova P, Hegde G. Recent trends and progress in corrosion inhibitors and electrochemical evaluation. Applied Sciences. 2023; **13**(18):10107. DOI: 10.3390/app131810107

[226] Faccini M, Bautista L, Soldi L, Escobar AM, Altavilla M, Calvet M, et al. Environmentally friendly anticorrosive polymeric coatings. NATO Advanced Science Institutes Series E: Applied Sciences. 2021;**11**:3446. DOI: 10.3390/app11083446

[227] Kozhukharov S, Girginov C, Kostadinova O. Highly ordered AAO layers for various sensor applications. In: Petkov P, Tsiulyanu D, Popov C, Kulisch W, editors. Advanced Nanotechnologies for Detection and Defence against CBRN Agents, NATO Science for Peace and Security Series B: Physics and Biophysics. Dordrecht: Springer; 2018. DOI: 10.1007/978-94-024-1298-7_43

[228] Knecht TA, Hutchison JE. Reaction atmospheres and surface ligation control surface reactivity and morphology of cerium oxide nanocrystals during continuous addition synthesis. Inorganic Chemistry. 2022;**61**(11):4690-4704. DOI: 10.1021/acs.inorgchem.1c03993

[229] Shah MAKY, Lu Y, Mushtaq N, Yousaf M, Zhu B. Unraveling the synergistic effect on ionic transport of ceria via the surface engineering for low-temperature ceramic fuel cells. Applied Physics Letters. 2023;**122**:113902. DOI: 10.1063/5.0137303

[230] Shah MAKY, Lu Y, Mushtaq N, Yousaf M, Zhu B. Doped ceria electrolyte rich in oxygen vacancies for boosting the fuel cell performance of LT-CFCs. International Journal of Hydrogen Energy. 2023;**48**(33): 12474-12484. DOI: 10.1016/j.ijhydene.2022.12.153

[231] Pundir S, Priya R, Singh K, Kaur H, Choudhary P. A systematic study on synthesis of CeO_2 nanoparticles by various routes. IOP Conference Series: Earth and Environmental Science. 2023; **1110**:012030. DOI: 10.1088/1755-1315/1110/1/012030

[232] Kumar P, Kumar P, Kumar A, Meena RC, Tomar R, Chand F, et al. Structural, morphological, electrical and dielectric properties of Mn doped CeO2. Journal of Alloys and Compounds. 2016; **672**:543-548. DOI: 10.1016/j.jallcom.2016.02.153

[233] Kabir A, Zhang H, Esposito V. Chapter 5 - Mass diffusion phenomena in cerium oxide. In: Scirè S, Palmisano L, editors. Metal Oxides, Cerium Oxide (CeO_2): Synthesis, Properties and Applications. Elsevier; 2020. pp. 169-210. ISBN 9780128156612. DOI: 10.1016/B978-0-12-815661-2.00005-0

[234] Al AN, Al-Reyahi AY, Alnemrat S, Twal AA. Stoichiometric and reduced ceria surfaces: atomic structure, energetics and electron localization. Computational & Theoretical Chemistry. 2023;**1225**:114157. DOI: 10.1016/j.comptc.2023.114157

[235] Schmitt R, Nenning A, Kraynis O, Korobko R, Frenkel AI, Lubomirsky I, et al. A review of defect structure and chemistry in ceria and its solid solutions. Chemical Society Reviews. 2020;**49**: 554-592. DOI: 10.1039/C9CS00588A

[236] Kilner JA. Defects and conductivity in ceria-based oxides. Chemistry Letters. 2008;**37**(10):1012-1015. DOI: 10.1246/cl.2008.1012

[237] Kozhukharov S, Nenova Z, Nenov T, Machkova M, Kozhukharov V. Influence of Ce(III)/Ce(IV) - supplements on the characteristics of humidity sensors with TiO_2 films prepared via a sol-gel method. Boletin de la Sociedad Espanola de Ceramica y Vidrio. 2013;**52**(2):71-78. DOI: 10.3989/cyv.92013

[238] Kozhukharov S, Nenova Z, Nenov T, Nedev N, Machkova M. Humidity sensing elements based on cerium doped titania-silica thin films prepared via a sol-gel method. Sensors and Actuators B: Chemical. 2015;**B-210**: 676-684. DOI: 10.1016/j.snb.2014.12.119

[239] Nenova Z, Kozhukharov S, Nenov T, Nedev N, Machkova M. Combined influence of titania and silica precursors on the properties of thin film humidity sensing elements prepared via a sol–gel method. Sensors and Actuators B Chemical. 2016;**B-224**:143-152. DOI: 10.1016/j.snb.2015.10.010

[240] Jiang X, Wang M, Hou L, Lin T. An ordered one-step colorimetric sensor for the selective determination of catechol based on the polyacrylic acid-coated cerium oxide with laccase-like activity. New Journal of Chemistry. 2022;**46**(46): 22412-22418. DOI: 10.1039/D2NJ04149A

[241] Tananaiko O, Walcarius A. Composite silica-based films as platforms for electrochemical sensors. The Chemical Record. 2023:e202300194. DOI: 10.1002/tcr.202300194

[242] Seminko V, Maksimchuk P, Klochkov V, Neuhodov Y, Demchenko L, Yefimova S. Reversible CeO_{2-x} and $CeO_{2-x}:Eu^{3+}$ Luminescent Hydrogen Peroxide Sensors with Recovery Rates Controlled by Temperature and UV Irradiation. The Journal of Physical Chemistry. 2023;**C-127**(22):10662-10669. DOI: 10.1021/acs.jpcc.3c02637

[243] Nenova Z, Nenov T, Kozhukharov S, Machkova M. Humidity Sensing Elements Based on Zn-doped TiO_2 Films Prepared via a Sol-gel Method. Materials Science (Medžiagotyra). 2015;**21**(2):191-195. DOI: 10.1016/j.jscs.2018.01.005

[244] Islam S, Bakhtiar H, Bidin N, Salim AA, Riaz S, Abbas KN, et al. Influence of ZnO doping on structural, optical and pH-stimulus characteristics of silica-titania nanocomposite matrix. Journal of Saudi Chemical Society. 2018; **22**(7):826-837. DOI: 10.1016/j.jscs.2018.01.005

[245] Islam S. Au and zincite assisted silica-titania nanocomposite coated fiber optic pH sensor: Structural thermal, and optical characteristics. Optical Materials. 2023;**136**:113405. DOI: 10.1016/j.optmat.2022.113405

[246] Ahmad W, Basit MA, Khan MS, Ali I, Park TJ. Superior atomic layer deposition of conformal ZnO shell on spherical SiO_2 particles for enhanced photocatalytic activity. Physica. 2020; **124**:114308. DOI: 10.1016/j.physe.2020.114308

[247] Islam S, Bakhtiar H, Haider Z, Alshoaibi A, Riaz S, Naseem S. Silica-titania nanocomposite based fiber optic sensor for aromatic hydrocarbons detection. Optics Communications. 2020;**471**:125825. DOI: 10.1016/j.optcom.2020.125825

[248] da Silva PAM, Rocha LSR, Desimone PM, Giulietti G, Aldao CM, Longo E, et al. Effect of thermal treatment on the 4f-hopping conductivity of CeO_2 exposed to CO(g) atmosphere. Materials Science and Engineering. 2023;**292**:116403. DOI: 10.1016/j.mseb.2023.116403

[249] Suzuki K, Kato M, Sunaoshi T, Uno H, Carvajal-Nunez U, Nelson AT, et al. Thermal and mechanical properties of CeO_2. Journal of the American Ceramic Society. 2019;**102**(4): 1994-2008. DOI: 10.1111/jace.16055

[250] Malakkal L, Prasad A, Oladimeji D, Jossou E, Ranasinghe J, Szpunar B, et al.

Atomistic and experimental study on thermal conductivity of bulk and porous cerium dioxide. Scientific Reports. 2019; **19**(9–1):6326. DOI: 10.1038/s41598-019-42807-5

[251] Celik E, Cop P, Negi RS, Mazilkin A, Ma Y, Klement P, et al. Design of ordered mesoporous CeO_2–YSZ nanocomposite thin films with mixed ionic/electronic conductivity via surface engineering. ACS Nano. 2022; **16**(2):3182-3193. DOI: 10.1021/acsnano.1c11032

[252] Gupta M, Shirbhate SC, Rambadey OV, Acharya SA, Sagdeo PR. Temperature-dependent delocalization of oxygen vacancies in La-substituted CeO_2. ACS Applied Energy Materials. 2022;**5**(8):9759-9769. DOI: 10.1021/acsaem.2c01442

[253] Kaur T, Singh K, Kolte J. Effect of intrinsic and extrinsic oxygen vacancies on the conductivity of Gd-doped CeO_2 synthesized by a sonochemical route. The Journal of Physical Chemistry. 2022; **C-126**(42):18018-18028. DOI: 10.1021/acs.jpcc.2c04851

[254] Yang G, El Loubani M, Chalaki HR, Kim J, Keum JK, Rouleau CM, et al. Tuning ionic conductivity in fluorite Gd-doped CeO_2-Bixbyite RE_2O_3 (RE = Y and Sm) multilayer thin films by controlling interfacial strain. ACS Applied Electronic Materials. 2023;**5**(8):4556-4563. DOI: 10.1021/acsaelm.3c00724

[255] Kaur T, Kumar R, Khan S, Singh K. Structural and Electrical Properties of Gd Doped CeO_2 (GDC) Nanoceramics for Solid Oxide Fuel Cell Applications. Transactions of the Indian Ceramic Society. 2022;**81**(3):127-132. DOI: 10.1080/0371750X.2022.2115400

[256] Kurmangaleev KS, Ikim MI, Kozhushner MA, Trakhtenberg LI. Electron distribution and electrical resistance in nanostructured mixed oxides CeO_2-In_2O_3. Applied Surface Science. 2021;**546**(30):149011. DOI: 10.1016/j.apsusc.2021.149011

[257] Kurapova OY, Glukharev AG, Glumov OV, Konakov VG. The effect of the sintering parameters on the structure and oxygen ion conductivity of Y_2O_3–ZrO_2–CeO_2 ceramics. Open Ceramics. 2021;**5**:100086. DOI: 10.1016/j.oceram.2021.100086

[258] Vishwakarma MK, Jain P. Investigation of temperature-activated charge-carrier dynamics and dielectric relaxation in Fe, Cu–Doped CeO_2. Physica Status Solidi. 2023;**220**(19): 230049. DOI: 10.1002/pssa.202300349

[259] Bosigo R, Lepodise LM, Kuvarega A, Muiva C. Hydrothermal synthesis of CuO and CeO_2/CuO nanostructures: Spectroscopic and temperature dependent electrical properties. Journal of Materials Science: Materials in Electronics. 2021;**32**: 7136-7152. DOI: 10.1007/s10854-021-05423-6

[260] Koyale PA, Delekar SD. A review on practical aspects of CeO_2 and its composites for photoelectrochemical water splitting. International Journal of Hydrogen Energy. 2024;**51A**:515-530. DOI: 10.1016/j.ijhydene.2023.06.251

[261] Sk R, Aziz SKT, Kumar S, Nessim GD, Ghosh K. 3D-graphene decorated with g-C_3N_4/Cu_3P composite: A noble metal-free bifunctional electrocatalyst for overall water splitting. ChemCatChem. 2020;**12**(5):1394-1402. DOI: 10.1002/cctc.201902065

[262] Rashid J, Parveen N, ul Haq T, Iqbal A, Talib SH, Awan SU, et al. g-C_3N_4/CeO_2/Fe_3O_4 ternary composite as an efficient bifunctional catalyst for

overall water splitting. ChemCatChem. 2018;**10**(24):5587-5592. DOI: 10.1002/cctc.201801597

[263] Du L, Gao XY, Wang GY, Yang CC, Jiang Q. CeO_2 nanoparticles decorated on porous Ni–Fe bimetallic phosphide nanosheets for high-efficient overall water splitting. Materials Research Letters. 2023;**10**(2):159-167. DOI: 10.1080/21663831.2022.2131372

[264] Zhang X, Liu F, Ji X, Cui L, Li C, Liu J. Facile generation of CeO_2 nanoparticles on multiphased NiS_x nanoplatelet arrays as a free-standing electrode for efficient overall water splitting. Journal of Colloid and Interface Science. 2024;**A-653**:308-315. DOI: 10.1016/j.jcis.2023.09.016

[265] Du X, Ding Y, Zhang X. Selectively Se-doped Co_3O_4-CeO_2 nanoparticle-dotted nanoneedle arrays for high-efficiency overall water splitting. Applied Surface Science. 2021;**562**: 150227. DOI: 10.1016/j.apsusc.2021.150227

[266] Chen H, Huang H-B, Li H-H, Zhao S-Z, Wang L-D, Zhang J, et al. Self-supporting Co/CeO_2 heterostructures for ampere-level current density alkaline water electrolysis. Inorganic Chemistry. 2023;**62**(7):3297-3304. DOI: 10.1021/acs.inorgchem.2c04525

[267] Song X-Z, Zhu W-Y, Ni J-C, Zhao Y-H, Zhang T, Tan Z, et al. Boosting hydrogen evolution electrocatalysis via regulating the electronic structure in a crystalline–amorphous CoP/CeO_x p–n heterojunction. ACS Applied Materials & Interfaces. 2022;**14**(29):33151-33160. DOI: 10.1021/acsami.2c06439

[268] Guo X, Li M, Qiu L, Tian F, He L, Geng S, et al. Engineering electron redistribution of bimetallic phosphates with CeO_2 enables high-performance overall water splitting. Chemical Engineering Journal. 2023;**453**(1): 139796. DOI: 10.1016/j.cej.2022.139796

[269] Zhang L, Lei Y, Xu W, Wang D, Zhao Y, Chen W, et al. Highly active and durable nitrogen-doped CoP/CeO_2 nanowire heterostructures for overall water splitting. Chemical Engineering Journal. 2023;**460**:141119. DOI: 10.1016/j.cej.2022.141119

[270] Ma G, Du X, Zhang X. Selective sulfuration, phosphorization and selenylation: A universal strategy toward Co-Ni-M-CeO_2/NF (M = O, S, P and Se) interface engineering for efficient water splitting electrocatalysis. Journal of Alloys and Compounds. 2021;**864**: 158486. DOI: 10.1016/j.jallcom.2020.158486

[271] Kozhukharov S, Tchaoushev S. Spray pyrolysis equipment for various applications. Journal of Chemical Technology and Metallurgy. 2013;**48**(1): 111-118

[272] Ho S, Ahmed N. An overview on spray pyrolysis deposition of metal oxide thin films. Research Journal of Chemistry and Environment. 2023; **27**(11):131-141. DOI: 10.25303/2711rjce1310141

[273] Touidjen NH, Lamrizeggar M, Aida MS, Rouabah S, Aouabdia N. Volatile organic compound gas sensing applications of n-type SnO_2 and p-Type CuO based on thin films. Journal of Electronic Materials. 2024;**53**(1): 515-524. DOI: 10.1007/s11664-023-10559-8

[274] Choubari MS, Mazloom J, Ghodsi FE. Supercapacitive properties, optical band gap, and photoluminescence of CeO_2–ZnO nanocomposites prepared by eco-friendly green and citrate sol-gel

methods: A comparative study. Ceramics International. 2022;**48**(15):21344-21354. DOI: 10.1016/j.ceramint.2022.04.100

[275] Das HT, Elango BT, Dutta S, Das N, Das P, Mondal A, et al. Recent trend of CeO_2-based nanocomposites electrode in supercapacitor: A review on energy storage applications. Journal of Energy Storage. 2022;**50**:104643. DOI: 10.1016/j.est.2022.104643

[276] Garcés L, López-Mireles DA, Padmasree KP, Mtz-Enriquez AI, Encinas A, Oliva J. Stretchable supercapacitors made with plasticine/graphene/$CeNdNiO_2$ composite electrodes. Materials Chemistry and Physics. 2024;**313**:128721. DOI: 10.1016/j.matchemphys.2023.128721

[277] Hunpratub S, Chullaphan T, Chumpolkulwong S, Chanlek N, Phokha S. Characterization and electrochemical properties of Carbon/CeO_2 composites prepared using a hydrothermal method. Materials Chemistry and Physics. 2023;**303**: 127820. DOI: 10.1016/j.matchemphys.2023.127820

[278] Qi H, Wang W, Yang N, Chen W, Yang X, Fang X, et al. Ultra-high cycling stability of 3D flower-like $Ce(COOH)_3$ for supercapacitor electrode via a facile and scalable strategy. Molecules. 2023; **28**(19):6806. DOI: 10.3390/molecules28196806

[279] Muñoz-Chilito JF, Rodríguez-Páez JE. ZnO-CeO_2 nanocomposites: Synthesis, characterization and evaluation of their action on polluting gases emitted by motorcycles. Journal of Environmental Chemical Engineering. 2021;**9**(1):104890. DOI: 10.1016/j.jece.2020.104890

[280] Huang Y, Liu Y, Deng Y, Zhang J, He B, Sun J, et al. Enhancing the bifunctional activity of $CoSe_2$ nanocubes by surface decoration of CeO_2 for advanced zinc-air batteries. Journal of Colloid and Interface Science. 2022;**625**: 839-849. DOI: 10.1016/j.jcis.2022.06.094

[281] Li T, He Z, Liu X, Jiang M, Liao Q, Ding R, et al. Interface interaction of Ag-CeO_2-Co_3O_4 facilitate ORR/OER activity for Zn-air battery. Surfaces and Interfaces. 2022;**33**:102270. DOI: 10.1016/j.surfin.2022.102270

[282] Piperno L, Vannozzi A, Augieri A, Masi A, Mancini A, Rufoloni A, et al. High-performance Fe(Se,Te) films on chemical CeO_2-based buffer layers. Scientific Reports. 2023;**13**:569. DOI: 10.1038/s41598-022-24044-5

[283] Manohar A, Reddy GR, Hatshan MR, Goud JP, Kim KH. Structural and magnetic properties of $Ca_{0.5}Mg_{0.5}Fe_2O_4/CeO_2/NiFe_2O_4$ nanocomposite for energy storage applications. Ceramics International. 2023;**49**(22):35392-35398. DOI: 10.1016/j.ceramint.2023.08.214

[284] Graça VCD, Holz LIV, Araújo AJM, Loureiro FJA, Fagg DP. A novel niobium (oxy)nitride-$BaCe_{0.7}Zr_{0.1}Y_{0.2}O_{3-\delta}$ composite electrode for proton ceramic membrane reactors (PCMRs). Journal of Energy Storage. 2023;**6815**:107769. DOI: 10.1016/j.est.2023.107769

[285] Saliyski N, Machkova M, Fagg D, Frade Z, Simeonov S. Freeze dried preparation of solid state electrolyte $Ce_{0.8}Re_{0.2}O_{1.9}$ (Re = Gd and Pr) materials. Annual Proceedings of the University of Rousse (Bulgaria). 2010; **49**:60-64. Available from: http://conf.uni-ruse.bg/bg/docs/cp10/9.1/9.1-10.pdf

[286] Kang W, Ozgur DO, Varma A. Solution combustion synthesis of high surface area CeO_2 nanopowders for

catalytic applications: Reaction mechanism and properties. ACS Applied Nano Materials. 2018;**1**(2):675-685. DOI: 10.1021/acsanm.7b00154

[287] Salian A, Akshay PK, Mandal S. Phase stabilized solution combustion processed $(Ce_{0.2}La_{0.2}Pr_{0.2}Sm_{0.2}Y_{0.2})O_{1.6-\delta}$: An exploration of the dielectric properties. Journal of Alloys and Compounds. 2023;**960**:170786. DOI: 10.1016/j.jallcom.2023.170786

[288] Manikanika LC, Rajesh K. Photocatalytic degradation efficiencies of ZnO nanoparticles and CeO_2 nanosheets synthesized via combustion method. Bulletin of Materials Science. 2023;**46**(3):181. DOI: 10.1007/s12034-023-03028-9

[289] Aguilar-Martínez O, Santolalla-Vargas CE, Tzompantzi F, González F, Romero-Ibarra IC, de los Reyes JA, et al. Improvement of the photocatalytic activity of CeO_2 on the degradation of chlorophenolic compounds obtained by solution combustion synthesis (SCS). Effect of urea as fuel. Solid State Sciences. 2023;**143**:107274. DOI: 10.1016/j.solidstatesciences.2023.107274